SURFACE WAVE ANALYSIS FOR NEAR SURFACE APPLICATIONS

It's not what you look at that matters, it's what you see.

Henry David Thoreau

SURFACE WAVE ANALYSIS FOR NEAR SURFACE APPLICATIONS

GIANCARLO DAL MORO

Institute of Rock Structure and Mechanics Academy of Sciences of the Czech Republic, Prague, Czech Republic
&
Eliosoft, Udine, Italy

ELSEVIER

AMSTERDAM   BOSTON   HEIDELBERG   LONDON   NEW YORK   OXFORD
PARIS   SAN DIEGO   SAN FRANCISCO   SINGAPORE   SYDNEY   TOKYO

Elsevier
Radarweg 29, PO Box 211, 1000 AE Amsterdam, Netherlands
The Boulevard, Langford Lane, Kidlington, Oxford OX5 1GB, UK
225 Wyman Street, Waltham, MA 02451, USA

Notices
Knowledge and best practice in this field are constantly changing. As new research and experience
broaden our understanding, changes in research methods, professional practices, or medical treatment
may become necessary.

Practitioners and researchers must always rely on their own experience and knowledge in evaluating and
using any information, methods, compounds, or experiments described herein. In using such information
or methods they should be mindful of their own safety and the safety of others, including parties for
whom they have a professional responsibility.

To the fullest extent of the law, neither the Publisher nor the authors, contributors, or editors, assume any
liability for any injury and/or damage to persons or property as a matter of products liability, negligence or
otherwise, or from any use or operation of any methods, products, instructions, or ideas contained in the
material herein.

Library of Congress Cataloging-in-Publication Data
Dal Moro, Giancarlo, 1969-
   Surface wave analysis for near surface applications / Giancarlo Dal Moro.
     pages cm
   Includes bibliographical references and index.
   ISBN 978-0-12-800770-9
   1. Surface waves (Seismology) 2. Seismology. I. Title.
   QE538.5.D35 2015
   551.22028'7–dc23
                                                         2014035516

British Library Cataloguing-in-Publication Data
A catalogue record for this book is available from the British Library.

ISBN: 978-0-12-800770-9

For information on all Elsevier publications
visit our website at http://store.elsevier.com

Working together
to grow libraries in
developing countries

www.elsevier.com • www.bookaid.org

CONTENTS

PREFACE

Learn the rules so you know how to break them properly

Dalai Lama

Rules are for fools

American folklore

In the mind of the Author, this book represents an attempt to fill the gap between academic research and field applications. Purely academic works sometimes risk to become self-referential efforts without reflecting in a genuine support for the society while, on the other side, practitioners and researchers facing the problem of field-data analysis without a sufficiently-robust theoretical background risk to fail because of erroneous comprehension of the data.

In fact, no methodology approach or software/hardware application will ever allow proceeding in sound data analysis without a clear understating of the considered phenomena by the side of the user.

Due to the several cross-references to different sections, the book should be considered in a holistic perspective (to some degree, the understanding of one point requires the understanding of all the others). Equations and formulas (easily found in the mentioned literature) were intentionally avoided while the focus was kept on the surface-wave phenomenology so to train the eye in the reading of the velocity spectrum (the fundamental *object* to consider in surface-wave analysis).

The principles characterizing the considered active and passive methodologies are necessarily presented without any claim or presumption to describe the whole universe of possible techniques and approaches.

Of course, as always in this kind of editorial projects, it is impossible to mention all the papers and studies that have been published on these subjects and we thus pay homage to the comprehension of all those researchers that, being not cited, will practice their patience and comprehension about that.

The overall goal is to provide an arsenal of conceptual and practical tools capable (when properly considered) to *attack* (as mathematicians often like to say) a site and *solve* it in terms of reconstruction of its subsurface model.

No easy and cheap approach based on simplistic assumptions can in fact result in universally valid analyses and no approach can be claimed as *the* ultimate one, suitable and sufficient for all the seasons. It will be shown that some rigid *rules* often proposed and adopted for surface-wave data acquisition and analysis can be inadequate or even fully misleading, thus realizing the concreteness of the wisdom pervading the opening quotes.

The *best* approach (and by the term *best* we can define the approach capable of unambiguously solving a site ensuring at the same time the minimum effort in terms of data acquisition and analysis) is in fact site-dependant and, in general terms, only the joint analysis of different (but related) data sets can guarantee the success of a survey.

Case studies presented in the Appendix represent a vital component of the book and a major effort was undertaken in order to select data sets capable of providing a sufficiently large outlook on the richness and complexity of the surface-wave phenomenology, approached and solved by adopting the methodologies that best suited the data themselves.

Giancarlo Dal Moro

CHAPTER 1

Introducing Surface Waves

We begin where we are.

Robert Fripp

1.1 A BRIEF INTRODUCTION

As very well known from basic seismology courses, fundamentally there are two kinds of seismic waves: those propagating inside a medium (*body waves*) and those traveling along the very shallow part of it (*surface waves (SWs)*). Compressional waves (commonly indicated as P waves) and shear waves (S waves) are body waves while Rayleigh, Scholte, Stoneley, and Love waves are different kinds of SWs.

In the last decades, a number of papers dealing with SWs have been published but it must be recalled that their theoretical description and first applications date back to almost a century ago.

SWs have been in fact used for a number of applications since the 1920s: Nondestructive testing (even for medical applications), geotechnical studies, and crustal seismology (e.g., Gutenberg, 1924; Evison et al., 1959; Viktorov., 1967; McMechan and Yedlin, 1981; Kovach, 1978; Roesset, 1998; Stokoe et al., 1988; Stokoe and Santamarina, 2000; Jørgensen and Kundu, 2002; O'Neill et al., 2003; 2004; Gaherty, 2004; Pedersen et al., 2006; Luo et al., 2007; O'Connell and Turner, 2011; Prodehl et al., 2013).

Recently the interest toward their application has increased both for the increasing demand for efficient methodologies to apply in geotechnical studies and because the recent regulations addressing the assessment of the seismic hazard (see for instance the *Eurocode8*) are giving the necessary emphasis to the determination of the shear-wave velocity vertical profile.

Because of their practical importance and wide use in a number of near-surface applications, we will focus our interest on Rayleigh and Love waves in the following.

1.2 LORD RAYLEIGH AND PROF. LOVE

There are two kinds of SWs actually relevant while analyzing seismic waves propagating on land: Rayleigh and Love waves. The first ones were described mathematically by Lord Rayleigh in 1885 (Rayleigh, 1885), while it was Prof. Love who, in 1911, described the kind of waves that were then named after him (Love, 1911).

The fundamental characteristics of Rayleigh waves are represented in the sketch reported in Figure 1.1. The wave (traveling in the direction of propagation) induces an

Surface Wave Analysis for Near Surface Applications
ISBN 978-0-12-800770-9, http://dx.doi.org/10.1016/B978-0-12-800770-9.00001-7

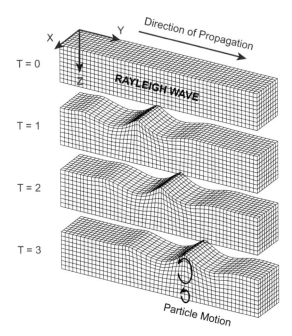

Figure 1.1 Rayleigh waves. *T* represents the time (the wave motion is depicted at three moments successive to the wave generation). The particle motion determined by the traveling Rayleigh wave occurs both on the vertical and horizontal planes (retrograde elliptical motion). On the horizontal plane the motion is along the radial component (see also Figures 1.2 and 1.3). *From http://www.geo. mtu.edu/UPSeis/waves.html.*

elliptical (retrograde) motion (see the blue ellipse drawn at time $T = 3$) whose amplitude exponentially decreases with depth. Such elliptical motion is the result of the superposition of the vertical and horizontal (more specifically radial) components (Figure 1.2).

Love waves are somehow simpler than Rayleigh waves because (Figures 1.3 and 1.4) they move only on the horizontal plane, transversally with respect to the direction of propagation. Incidentally, this *simplicity* also mirrors in both the computational load necessary to solve their constitutive equations (and describe their propagation), both in their phenomenology which, how we will broadly see in the next chapters and in several presented case studies, will result extremely useful (even necessary) to solve puzzling interpretative issues related to complex Rayleigh-wave velocity spectra.

Let us now summarize further basic facts:

- While considering a surface normal load, the energy converted into Rayleigh waves is by far predominant (67%) with respect to the energy that goes into P (7%) and S (26%) waves (Miller and Pursey, 1955);
- Rayleigh and Love waves are called SWs because their amplitude exponentially decreases with depth, thus the motion induced by their passage is limited to a shallow portion (whose depth depends on the considered wavelength λ—see later on);

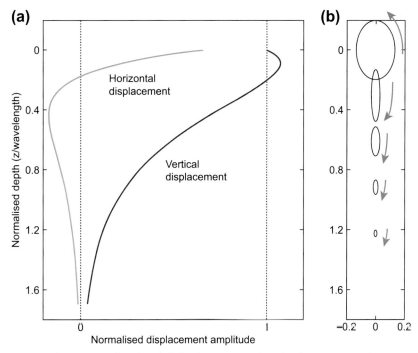

Figure 1.2 Normalized vertical and radial displacements of Rayleigh waves as a function of depth (normalized with respect to the considered wavelength): (a) the individual displacements of the vertical and radial components and (b) the elliptical motion resulting from the composition of the vertical and radial movements. *From Gedge and Hill (2012).*

- Just because their energy is confined to a shallow layer, while expanding from the source (*geometrical spreading*), their amplitude decreases fundamentally according to the square root of the distance from the source, while body waves (whose propagation involves a semisphere and not just a circle) lose their energy (thus amplitude) according to the distance (because of this, the amplitude of the body waves decreases much more with respect to SWs and consequently SWs tend to dominate the data);
- Compared to body waves, their amplitude is remarkably larger and, for this reason, in the low-frequency range they dominate the data and are therefore often referred to as *ground roll* (Figure 1.5 reports a classical common-shot gather giving evidence of this);
- Rayleigh waves move along a radial plane (they have both a radial and vertical component) according to a *retrograde* movement (that means that the elliptical particle motion is on the opposite direction with respect to the direction of propagation—see Figures 1.1, 1.2 and 1.4); Love waves (Figures 1.3 and 1.4) move only on the horizontal plane, with the particle motion perpendicular to the direction of propagation. The fact that Rayleigh waves have both a vertical and a horizontal component means that they can be acquired in a so-to-speak alternative way with respect the common

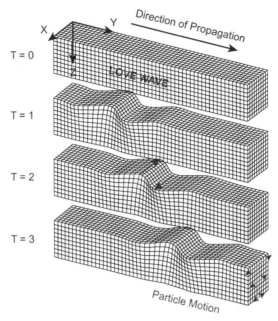

Figure 1.3 Love waves. *T* represents the time (the wave motion is depicted at three moments successive to the wave generation). The particle motion determined by the traveling Love wave lies only on the horizontal plane, transversally (i.e., perpendicularly) to the direction of propagation (see also Figure 1.3). *From http://www.geo.mtu.edu/UPSeis/waves.html.*

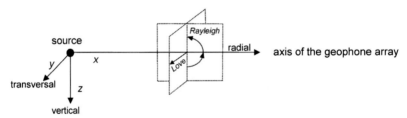

Figure 1.4 Ground motion associated to Rayleigh and Love waves: Rayleigh waves induce a motion along the vertical and radial axes, while Love waves along the transversal one.

practice represented by the use of vertical geophones: using horizontal geophones oriented radially with respect to the source (for further details see next Chapter). This can have a relevant series of theoretical and practical consequences that, in the following chapters and case studies, will be described in some detail.

Land acquisition is surely the most common, but what happens while considering marine (or lacustrine) seismic data (i.e., data traveling at a solid—fluid interface)?

While the characteristics of Love waves remain the same (e.g., Winsborrow et al., 2003), the so-to-speak "marine Rayleigh waves" are following slightly different equations that describe the so-called Scholte waves (Scholte, 1947).

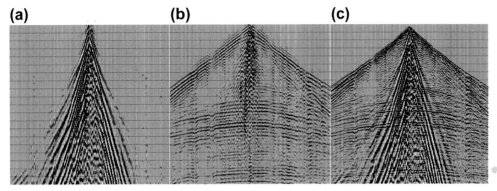

Figure 1.5 Example of common-shot gather containing both *ground roll* and reflections/refractions: (a) filtered from 0 to 15 Hz; (b) from 15 to 30 Hz; and (c) unfiltered. *(From Cary and Zhang (2009).)* Please notice that in the low-frequency range (0—15 Hz) the dataset is largely dominated by the *ground roll* (Rayleigh waves). On the other side, in the 15—30 Hz frequency range (high frequencies), data are dominated by refractions and reflections.

Scholte waves are actually quite similar to Rayleigh waves. The particle motion is absolutely analog (an elliptical motion on the radial—vertical plane) but, because of the influence of the water, the velocities are slightly different (Scholte waves tend to be slower). The difference between Rayleigh and Scholte waves results proportional to the thickness of the water column so that, from the practical point of view, in shallow waters to some degree it is possible to analyze Scholte waves while using the Rayleigh-wave equations.

Analyzing marine datasets, some difficulties can actually depend on guided waves traveling within the water column (e.g., Klein et al., 2005) and, depending on the specific goals of the survey, it can be desirable to use multicomponent sensors deployed on (or close to) the sea floor (e.g., Ritzwoller and Lavshin, 2003) rather than single-component hydrophones floating on the water column.

Stoneley waves are a further type of SW that create along a solid—solid interface and which are often exploited in borehole seismics to infer the shear-wave velocities (e.g., Stevens and Day, 1986).

1.3 DISPERSION FOR DUMMIES

It is well known that a seismic wavelet (actually any signal) is the result of several components (frequencies) that, all together, create the specific wavelet which can be described in terms of amplitude and phase spectra (any elementary signal processing textbook widely treats Fourier analysis and related topics).

The crucial point about SW propagation is that the propagation of a specific component (that is frequency) that compose the traveling seismic wavelet, depends on the medium properties from the surface down to a depth which is proportional to the wavelength of that specific component.

The very well-known fundamental relationship describing the propagation of an oscillation and that consequently depicts the link between wavelength λ, frequency f, and propagation velocity v is:

$$\lambda = v/f \tag{1.1}$$

An example will help understanding few basic and simple facts that will eventually result quite helpful in understanding/reading the velocity spectra of our field datasets.

Let us consider a 10 Hz component traveling along two different media: in one case a very soft sediment cover (for instance, saturate sand) and, in a second case, a conglomerate.

In the first case, a possible value for V_S is 100 m/s while in the second case (the conglomerate) the shear-wave velocity might be around 600 m/s. The wavelengths of the 10 Hz component will be 10 m (100/10) in the first case while 60 m (600/10) in the second case, where the same component is traveling over a conglomerate.

The same component (10 Hz) has thus a different wavelength depending on the material along which the wave travels: the higher the V_S of the material, the longer the wavelength.

The crucial point to consider when dealing with SWs is that the portion of subsurface that influence the propagation of a given component (both its velocity and amplitude) is more or less equal to $\lambda/3$ (such a simple rule of thumb is often referred to as *steady state approximation* and some very optimistic authors prefer to think in terms of $\lambda/2$).

The consequence is that, since the seismic wavelet is made of a range of components (in near-surface geotechnical applications we are commonly dealing with the dispersion of SWs in the 4–50 Hz frequency range), it will be possible to describe the medium down to a depth which depends on the largest wavelengths that is possible to consider with sufficient precision.

Figure 1.6 schematically illustrates this very basic concept: the low-frequency components sample the subsurface in depth while the high-frequency components are influenced only by the shallowest layers.

An elemental representation of this is possible and convenient in a frequency–velocity graph capable of summarizing the velocities of each frequency (i.e., component). For merely educational purposes, Figure 1.7 qualitatively schematizes two ideal cases that should help in fixing the basic point (the lower the frequency, the thicker the investigated portion). As a matter of fact, later on we will see that, since SWs can travel according to different modes, things can actually be by far much more complex.

Figure 1.7(a) illustrates a very simple subsurface model that, since the V_S is monotonically increasing with depth, produces a dispersion curve like the one reported on the right panel (phase velocities increase while considering lower and lower frequencies).

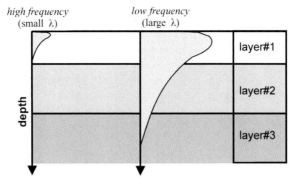

Figure 1.6 Schematic representation of two components that, because of the different wavelength, sense different subsurface portions: the high-frequency (i.e., small wavelength) component is influenced only by the shallowest layer(s) while the low-frequency component (large wavelength) travels at a speed that depends also on the deep layers.

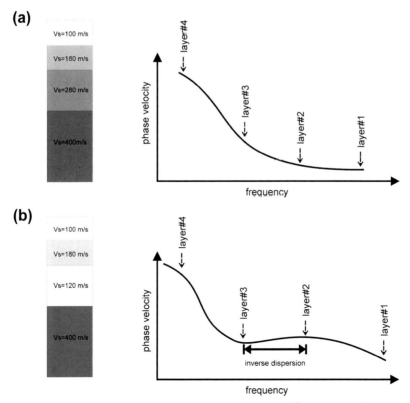

Figure 1.7 An elemental and rudimentary representation: two different subsurface models, two different phase-velocity dispersion curves showing (a) *normal* and (b) *inverse dispersion*.

Since, typically, deeper materials are more "compact" than the superficial ones, this sort of dispersion (to further understand the concept of *dispersion* see next paragraph) is often called *normal dispersion*.

On the other side, if a "softer" stratum is present in between two harder layers (Figure 1.7(b)) we can speak about low-velocity layer. In this case, in a certain frequency range (indicated by the line between the layer 2 and the layer 3) we can observe an *inverse dispersion*.

Considering the Eqn (1.1), it is clear that the largest wavelength is connected to the smallest frequency we can safely consider, that is the smallest frequency that we can clearly identify in the velocity spectrum and, consequently, include in our analyses.

If, for instance, we get back to the previously considered case (the conglomerate) and imagine that 10 Hz is the smallest frequency that we can safely consider (because, for several possible reasons, for lower frequencies the velocity spectrum appears blurred or lacking on any clear signal), in that case the maximum penetration depth will be about 20 m (60/3).

Seismologists are capable of imaging the whole crustal structure (several tens of kilometers) by exploiting very low-frequency components that are produced during large earthquakes and travel thousands of miles along the Earth surface (Gaherty, 2004; Pedersen et al., 2006; Prodehl et al., 2013).

It is anyway quite important to underline that such a rule of thumb can result quite problematic when dealing with complex velocity spectra for which the velocity spectra appear problematic to interpret (in the following chapters we will see how to choose the proper datasets/components and get as deep as possible while jointly considering datasets from active and passive acquisitions).

1.4 DISPERSION, VELOCITY SPECTRA, AND DISPERSION CURVES

Since in a layered medium different components travel at different speeds, the consequence is that the signal (i.e., the seismic wavelet) will undergo a dispersion process.

The literature where this process is mathematically described and quantified is immense (see References) but to make this concept clear to all those who are not interested in understanding this fact in mathematical terms, we can represent such a phenomenon through a simple analogy.

Figure 1.8 shows the typical military formation of the ancient Roman army (the *testudo* or *tortoise*) at two different moments (t1 and t2). At the time t1 (the moment in which the *testudo* starts marching toward the enemy) the formation is compact (i.e., the soldiers are regularly spaced and properly packed together).

In our analogy, each soldier can be imagined as a component (i.e., a specific frequency). At the very beginning (the moment t1) all the components are compacted in the *testudo* (which represents our source seismic wavelet).

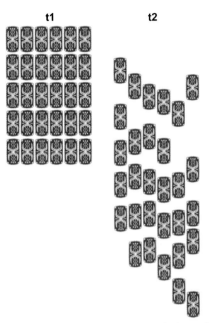

Figure 1.8 The concept of *dispersion* through the representation of a Roman *testudo* depicted (bird's-eye view) at two successive moments (t2 > t1): since each soldier moves at different speeds, the original *testudo* disperses.

Quite clearly, if all the components (soldiers) would march (travel) at the same velocity, the *testudo* would keep its original (regular) shape (and no *dispersion* would occur).

On the other side, if each soldier would move on his own "spontaneous" pace regardless of what the companions do, after a while the *testudo* would lose its original compactness and shape, and end up with a dispersed *testudo* (see the formation presented at time t2).

We can now kindly move to the seismic case. In Figure 1.9 (upper panel) we report an example of multichannel seismic dataset where the dispersion is quite evident: at the near-source trace the wavelet is "small" (i.e., compacted) because all the components are initially packed together. Because of the dispersion (some components run faster than others), after some *time* (which clearly means also *space*) the original seismic wavelet has enlarged (see its temporal length at the faraway offsets).

Since different authors can use slightly different terminologies, it is now important to clearly define the expression *velocity spectrum* and differentiate it with respect to the concept of *dispersion curve*.

In this book, analogously to most of the authors, by *velocity spectrum* we mean the *matrix* representing the propagation velocity as a function of the frequency (Figure 1.9, lower panel). The values in this matrix (thus the colors) represent a sort of "correlation factor"

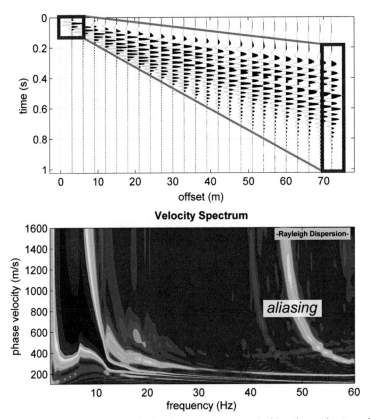

Figure 1.9 Dispersion of a seismic signal. The dispersion is revealed by the widening of the seismic wavelet associated to the considered surface wave (see red rectangles in the upper panel) that, traveling along the array modifies its characteristics (see the difference between the first and the last traces). By transforming the seismic data from the offset–time domain (upper panel) into the frequency–velocity domain (lower panel), we obtain the so-called velocity spectrum (Park et al., 1998; Dal Moro et al., 2003). Spatial aliasing will be illustrated and commented in the next chapter.

for each *frequency–velocity* point (in this case red means high correlation). Such a matrix is obtained by transforming the seismic traces (originally in the *offset–time* domain—Figure 1.9, upper panel) into the *frequency–velocity* domain (Park et al., 1998; Dal Moro et al., 2003). This means that a *velocity spectrum* is an objective "entity" mathematically derived from the seismic traces without any sort of interpretation from the user.

It is definitely essential to stress the attention on the fact that the velocity spectrum is the primary "object" to consider in SW analysis. All the other possible "objects" are fundamentally derived from it and require a deep understanding of a number of things that we can synthetically define as "surface wave phenomenology" (see Chapter 3).

On the opposite, a *dispersion curve* is a curve (a set of *frequency–velocity* points) summarizing the SW dispersive properties.

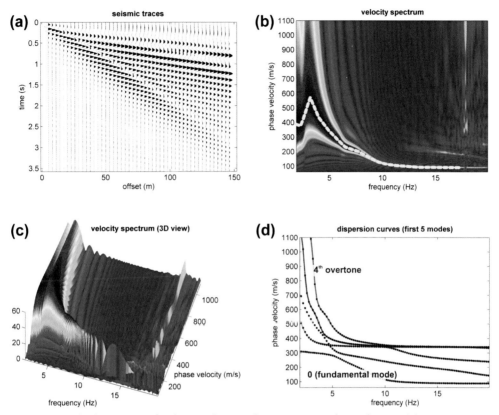

Figure 1.10 Clarifying some fundamental terms by using a synthetic dataset: (a) seismic traces; (b) velocity spectrum; (c) 3D view of the velocity spectrum (z-axis represents the correlation value for a given frequency—velocity point); (d) modal dispersion curves (indicated the forth higher mode and the fundamental one). The yellow curve (panel b) depicts the peaks in the velocity spectrum and is commonly referred to as *effective dispersion curve*.

Figure 1.10 will help defining some details about it. There are in fact two kinds of dispersion curves.

Commonly, most of the authors by *dispersion curve*(s) mean the modal dispersion curve(s) (see Figure 1.10(d)). These can actually come from the *interpretation* of a *velocity spectrum* in terms of modal dispersion curve(s) (nowadays this is unfortunately the standard way to approach SW analysis) or, on the opposite, can represent the theoretical dispersion curve(s) of a given subsurface model. This latter case is considered in Figure 1.10 where we are considering a synthetic dataset: given a model (i.e., a set of V_S, V_P, density, and thickness values describing the subsurface stratigraphy) it is in fact possible to determine its dispersive properties in terms of *modal* or *effective* dispersion curve(s) (e.g., Dunkin, 1965; Tokimatsu et al., 1992).

It is surely important to recall that, in order to fully describe the phenomenon, we should also consider the effects of the attenuation, thus the Q_S and Q_P quality factors (e.g., Panza, 1989; Lai and Rix, 2002; Lunedei and Albarello, 2009).

SWs represent a traveling vibration and, as any vibration, can have different modes (harmonics) which travel at different velocities (higher modes are faster—see Figure 1.10(d)). The modal dispersion curves represent the frequency—velocity curves associated to each mode (also called *overtone*).

Essentially, the *effective dispersion curve* is the "apparent" dispersion curve resulting from the superposition of all the modes (Tokimatsu et al., 1992) and, on a *velocity spectrum*, is thus represented by the peaks (see yellow dotted curve in Figure 1.10(b)). We will discuss this point in Chapter 6 and in some of the case studies illustrated in the Appendix.

Here it is important to emphasize that the *velocity spectrum* is something objective, mathematically obtained from the seismic traces without any personal interpretation. On the opposite side, when a *velocity spectrum* is interpreted in terms of *modal* dispersion curve(s), we are creating an "object" (a dispersion curve) which is clearly and necessarily a subjective interpretation (different practitioners can *pick* different curves according to their personal judgment).

In other words, inverting a picked modal dispersion curve does not mean to invert an observed (objective) data but rather a personal interpretation of it: it is absolutely fundamental to clearly understand this difference because an erroneous picking necessarily leads to meaningless results (e.g., Zhang and Chan, 2003).

Finally, if we pick the peak values of a velocity spectrum (see for instance the yellow dotted line in Figure 1.10(b)), we are dealing with an *effective* dispersion curve (which, to a large extent, is again an objective entity—see also Chapter 6) that anyway must be considered in a completely different way with respect to a *modal* dispersion curve.

The entire book is actually aimed at understanding how dangerous can be the simplistic approach based on the analysis of the *modal* dispersion curve(s) (that, we repeat that, nowadays represents a very common practice): if we are dealing with just one "object" (which is typically the velocity spectrum of the vertical component of Rayleigh waves), the risk of misinterpretation of the data in terms of *modal* dispersion curves is extremely high and only the joint analysis with further "objects" (for instance, Love waves) and/or an approach not based on the *effective* dispersion curve can save from erroneous analyses eventually necessarily producing meaningless subsurface reconstructions.

It is well known (e.g., Xia et al., 1999) that the parameters that mostly influence V_R (the propagation velocities of the Rayleigh waves) are the shear-wave velocity and the thickness of the layers. Density and V_P play a very minor role that can be easily quantified (and compared) by means of their partial derivatives. Therefore, V_P values and densities cannot be (seriously) retrieved from the analysis of the SW propagation (some authors tend, for instance, to fix some Poisson values and, after having identified the V_S values,

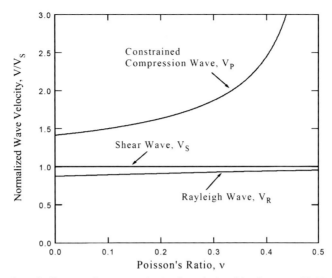

Figure 1.11 A uniform half space: the curves depict the relationships between V_S, V_P, and V_R velocities and Poisson's ratio. *(From Richart et al. (1970).)* Please notice that V_R is quite close to V_S and that the dependence on Poisson (i.e., V_P) is practically negligible.

consequently provide V_P values as well, but this is quite clearly a sort of trick rather than the result of a properly constrained analysis).

On the other side, Love wave propagation depends only on V_S, thickness, and, to a very minor extent, density (being V_P completely irrelevant).

The relationships between V_S, V_P, and V_R (as a function of the Poisson's ratio) for a uniform material are summarized in Figure 1.11. Two facts are straightforward:

1. V_R does not significantly depend on the Poisson's modulus (which is another way to underline that V_P has a practically negligible role);
2. V_R is slightly lower than V_S ($V_R \approx 0.9\ V_S$).

What happens when the medium is completely homogenous? In this case, Love waves cannot form (they show up only when the medium is layered), while Rayleigh waves propagate keeping a very simple aspect. Figure 1.12 reports a field dataset acquired in an area characterized by a thick sandy cover. It is quite apparent that the seismic wavelet does not modify significantly along the considered offsets (which is already evidence that very little dispersion takes place).

In this undemanding case (very little dispersion), computing the V_S is actually extremely easy even just from the seismic traces alone and simply considering the slope of the recorded Rayleigh wave. The indicated Δt and Δx quantities (Figure 1.12, left panel) are in fact the only quantities we need to compute V_R and then V_S. V_R is quite simply provided by $\Delta x/\Delta t$ (which in this case results about 140 m/s) while, considering the curves reported in Figure 1.11, V_S can be estimated simply by multiplying such a value by 1.1 ($V_S \approx 150-160$ m/s).

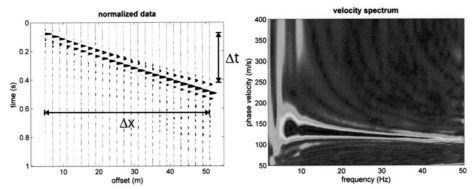

Figure 1.12 Real dataset (Rayleigh waves recorded by using vertical geophones and vertical-impact sledgehammer) showing almost no dispersion (evidence of quite homogenous subsurface stratigraphy). The seismic wavelet that travels along the array (left panel) does not significantly modify and the signal on the velocity spectrum (right panel) is almost flat (the phase velocity in the 5–50 Hz frequency range is more or less constant). Compare with seismic traces considered in Figures 1.9 and 1.10, where a layered model with large V_S variations produces a significant dispersion (while traveling the seismic wavelet widens).

If we give a look at the computed velocity spectrum (on the right panel), it is furthermore clear that the phase velocities are more or less constant all over a wide (5–50 Hz) frequency range (for lower frequencies, the used source—a common sledgehammer—did not produce enough energy). It will not be surprising to notice that the identified phase velocities are centered around an average value of about 140 m/s. Of course some dispersion is actually present (very high frequencies are slightly slower) but here the goal was to get familiar with basic aspects and to understand the link between the seismic traces (the data in the *offset–time* domain) and the velocity spectrum (data in the *frequency–velocity* domain), even if, for more complex datasets, the link could be definitely trickier.

1.5 ATTENUATION IN SHORT

For practical reasons, we are typically used to deal with normalized traces but the real amplitude of a seismic trace decreases with the offset. Trace normalization is a trivial operation (the values of each single trace are divided by the maximum amplitude of that specific trace so that the largest amplitude of each single trace becomes 1).

Anyway the amplitude of a seismic signal is subject to attenuation (e.g., Barton, 2006) mainly because of two reasons:

1. The so-called *geometrical spreading* (the energy spreads over a larger volume so that the energy density decreases and with it the amplitude of the wavelet);

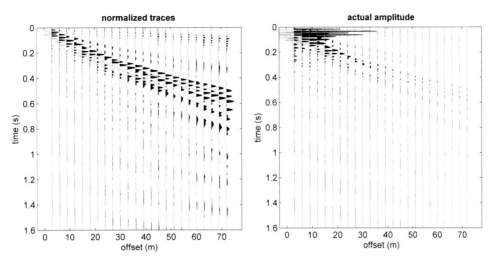

Figure 1.13 Normalized and actual seismic traces for a common-shot gather (24 vertical-component geophones and vertical-impact sledgehammer).

2. The *intrinsic* attenuation that determines a loss of energy proportional to the considered frequency (in general, since at each cycle some energy is lost, high frequencies undergo higher intrinsic attenuation).

To complete the scenario we must then surely recall that a further possible source of attenuation (that clearly determines a further decrease of the amplitude) is determined by scattering phenomena (e.g., Tselentis, 1998).

Figure 1.13 reports both real and normalized traces for a typical common-shot gather considered in near-surface applications (24 traces along a 72 m array). The large-amplitude Rayleigh waves arrive at the distant trace mostly between 0.4 and 0.9 s (since several higher modes are present and significant dispersion takes place, it is not possible to define too strictly the temporal window containing the Rayleigh-wave arrivals). A very well-defined refraction is also present (for details, see case study 2).

In Figure 1.14 is presented a synthetic dataset used for illustrating some basic facts regarding the attenuation but which is also extremely useful for emphasizing some problems connected to the difference between *modal* and *effective* dispersion curves (see previous paragraph). The quality factors Q_S were fixed proportionally to V_S values and following a simple rule of thumb: $Q_S = V_S/8$ ($Q_P = Q_S$).

Giving a look at the computed velocity spectrum and noticing the relationships with the overlaying *modal* dispersion curves (Figure 1.14(c)), it can be noticed that the *effective* dispersion curve (i.e., the peaks of the velocity spectrum) is dominated by the fundamental mode for frequencies higher than 6 Hz but by the first, second, and third higher modes in the 3−6 Hz frequency range. This example should immediately make clear two

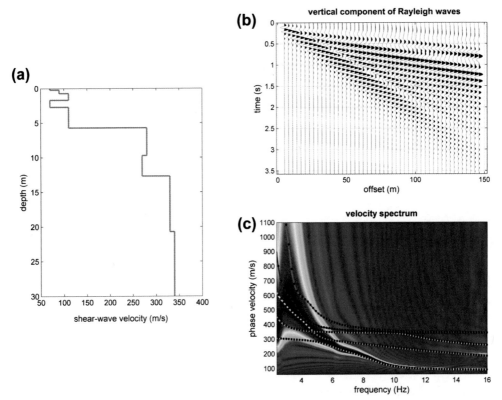

Figure 1.14 Rayleigh-wave dispersion for a synthetic dataset: (a) considered V_S profile; (b) normalized synthetic traces (vertical component of Rayleigh waves); (c) velocity spectrum with, overlaying, the modal dispersion curves of the first five modes.

remarkable facts that will be widely treated in the following chapters (and in the case studies presented in the Appendix):

1. The fundamental mode is not necessarily the dominating one.
2. The continuity of a signal in the velocity spectrum (i.e., the continuity of the *effective* dispersion curve) does not necessarily mean that it pertains to a single mode.

In fact, in this simple case the signal in the velocity spectrum (the red "curve" in Figure 1.14(c)) is a mix of different modes (the fundamental one for the high frequencies, and a complex mix of higher modes in the low-frequency range).

The successive Figure 1.15 reports some basic quantities necessary to describe SW attenuation. The *attenuation curves* of the field data (which can be thought in a similar way as the *dispersion* curves and that represent the attenuation as a function of frequency) can be determined according to different approaches (e.g., Tonn, 1991; White, 1992; Rix et al., 2000; Xia et al., 2002).

In the classical approach, their inversion (aimed at estimating the quality factors Q, in particular Q_S) can then be performed only after the determination of the V_S profile (Xia et al., 2002).

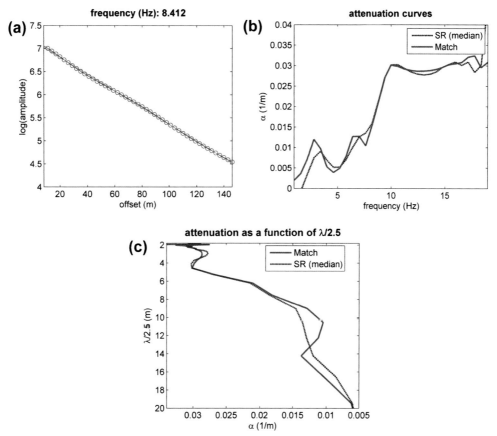

Figure 1.15 Rayleigh-wave attenuation for the synthetic model presented in Figure 1.14: (a) logarithm of the amplitude versus the offset for the 8.4 Hz frequency; (b) attenuation curves computed according to the "spectral ratio" and "matching" techniques (Tonn, 1991); (c) same attenuation curves plotted here as a function of $\lambda/2.5$ (following the *steady state approximation* to have an initial guess about the quality factors in depth); please notice the similarity with the V_s profile reported in Figure 1.14a and consider that the Q_s values were fixed according to the simple $Q_s = V_s/8$ rule of thumb.

Two aspects can anyway be problematic and, if not properly handled, lead to erroneous subsurface Q reconstruction:

1. Inadequate V_S model (which is a necessary preliminary input to analyze the attenuation) deriving from erroneous dispersion analysis (this is *the* crucial point that represents the main point of the entire book).
2. Coexistence of multiple modes with similar energy.

While the first point should not require any special explanation (it is quite intuitive that if one of the ingredients of the analysis is wrong, the entire process is going to suffer from that), for the second one we must recall that the basic relationship linking the α SW attenuation coefficient (see Xia et al., 2002) involves the partial derivatives of V_R with respect to V_S and V_P.

Since such derivatives are different for the different modes, in case of multiple modes present with similar energy, a problem arises: the common and simplistic assumption usually considered while analyzing the attenuation is in fact that a single mode (usually the fundamental one) is present, but this condition is actually quite rare in common field datasets.

Moreover, if different modes are dominating in different frequency ranges, the partial derivatives of each specific mode should be considered.

1.6 SURFACE WAVES, GEOLOGY, NONUNIQUENESS, AND ANISOTROPIES

In the previous paragraphs we have seen how the subsurface materials influence the propagation of the SWs. Since here we are basically interested in near-surface geotechnical applications, we can now start introducing typical values of V_S for common geological materials. In this respect Table 1.1 provides a very general overview on some classes of values (the variability is clearly so large that it would be completely useless to try to be more specific).

Very often, mixed materials such as the gravels can have very different V_S values (approximately ranging from 300 to 500 m/s) depending on the relative amount of cobblestones and fine matrix (usually made of sand or clay—see Figure 1.16).

In this respect it can be interesting to report the equation describing the resulting effective velocity when two different materials are mixed up (Stesky, 1978):

$$\frac{1}{V_m} = \phi \frac{1}{V_A} + (1 - \phi) \frac{1}{V_B} \tag{1.2}$$

where V_m is the velocity of the mixed media, V_A and V_B are the velocities of the materials A and B, and ϕ is the volume fraction of the material A in the mixed media.

Nonuniqueness of the solution is a problem for any surface methodology and we will discuss about this problem (and to its solution) in Chapter 5. Here the concept is briefly introduced because, although quite well known, very often the consequences are

Table 1.1 Typical Shear-Wave Velocities for Some Groups of Geological Materials Particularly Common in Near-Surface Applications

Material	V_S (m/s)
Peats and incompetent soils	50–150
Competent soils (silt, clay, sand)	150–350
Very competent soils and gravels	350–600
Weathered rock	600–800
Solid rock	>800

Figure 1.16 Gravel layer constituted by limestone cobblestones (high V_S) sparse in a silty matrix (low V_S). The resulting overall velocity value can be described by means of the Eqn (1.2).

neglected and very little care is paid to the ambiguity of the analyses provided while considering a single objective function.

Figure 1.17 introduces this point through the presentation of a series of V_S models and the related Rayleigh- and Love-wave dispersion curves for the first three modes.

Three features need to be strongly underlined:

1. In spite of the significant differences of the considered V_S model for depths higher than 8 m (Figure 1.17, left panel), the fundamental modes of all these models are practically identical (thus nonunique).
2. Higher modes are definitely more sensitive (they are more different than the fundamental one).
3. Down to a depth of about 8 m, the V_S model is definitely unique (so there is no ambiguity about the very shallow part).

The first point simply means that by using only the fundamental mode it is impossible to identify the real V_S structure for depths higher than about 8 m. Please notice that this value actually depends on the considered frequency range and cannot be considered universal, since investigated depth depends both on the available frequency range and the length of the array (seismologists can investigate the entire crust because they analyze very low frequencies that have traveled hundreds or thousands of kilometers). The frequency range considered in Figure 1.17 is anyway typical in most of the near-surface applications (4–40 Hz).

The reason why higher modes are more sensitive to deep V_S variations is basically related to the Eqn (1.1). Higher modes are in fact faster and, consequently, the corresponding wavelength is higher. This means that, for the same frequency, higher modes (if present) sense more intensely the deeper layers. Unfortunately, especially in the past when multimode inversion was still a problem, higher modes were often regarded as a problem but actually they bring much more information than the fundamental one and, if present, should be properly exploited.

Few words about the effects of possible anisotropies while analyzing SWs.

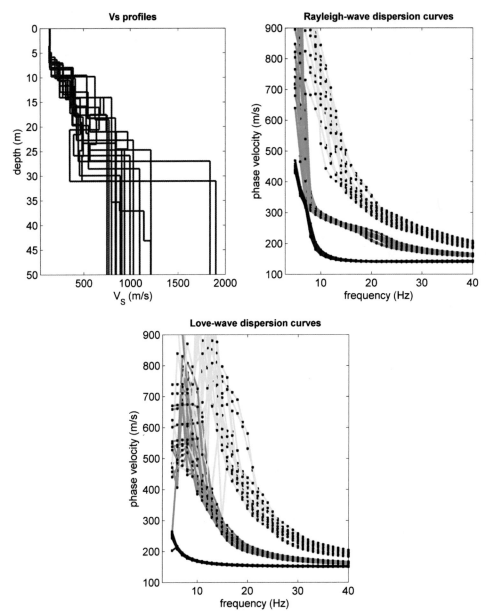

Figure 1.17 Elemental representation of nonuniqueness in surface-wave analysis: different subsurface V_S models can have very similar modal dispersion curves (in particular, low-order modes).

Stating that the propagation of Rayleigh and Love waves depends on V_S (Section 1.4) is actually a simplification valid only in case of homogenous media. When some anisotropy is present (Figure 1.18) V_S (and V_P) should be defined with respect to each axis. Since Rayleigh waves are influenced by the condition along the vertical axis, the

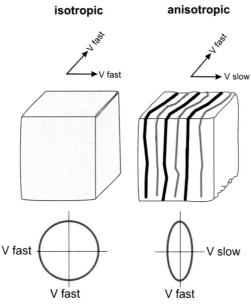

Figure 1.18 Seismic anisotropy induced by layers of cracked media.

pertinent V_S should then be indicated as V_{SV}. On the opposite, because of the kind of motion, Love-wave velocities are influenced by V_{SH}.

When dealing with highly layered materials such as for instance a *flysch*, it is quite intuitive that the velocities along the stratification and the one perpendicular to it can be significantly different. The seismic wave traveling perpendicular to the stratification is in fact forced to go through all the layers (also the soft ones) and, consequently, its velocity will be a weighted average depending of the velocities of each single layer (Figure 1.18, V *slow*).

On the opposite side, the wave traveling parallel to the stratification will somehow be free to travel along the "fast" layers and its velocity will be higher (Figure 1.18, V *fast*) compared to the one propagating perpendicularly to the layered medium.

If, for instance, we imagine a horizontally stratified rock formation, the anisotropy will be such that $V_{SH} > V_{SV}$.

In seismology, the anisotropy of relevant crustal and subcrustal layers is quite well known (see, for instance, Gaherty, 2004 and Pedersen et al., 2006) but very little studies are available with respect to near-surface materials (for a general overview, see also Barton, 2006).

A practical way to quantify the anisotropy (often indicated by the Greek letter ξ) is provided by the following relationship:

$$\xi = \frac{V_{SH} - V_{SV}}{V_{SH}} \cdot 100 \tag{1.3}$$

In this respect, when V_{SH} is higher than V_{SV} anisotropy is said to be *positive* (Gaherty, 2004).

Two final points are presented to underline about the anisotropy:

1. Considering the issues related to the nonuniqueness of the solution (see Figure 1.17 and related text) while considering the frequency range (4–40 Hz) typically used in near-surface applications, we can hope to identify possible anisotropies only in the very shallow layers (say down to about 8 m). To investigate in detail deeper layers, it is definitely necessary to go through sophisticate joint analyses that, in general terms, will be introduced in the next chapters.

2. In general, if anisotropy values up to, say, 10% might be reasonable, higher values are likely due to some problem/mistake in the data analysis.

In any case, the theoretical background and the experience required to face these aspects are such to recall the fundamental motto: *do it only if you (perfectly) know what you are doing.* Only if you fully master all the concepts and practical issues related to SW dispersion and attenuation you can hope to properly handle these subtle aspects.

CHAPTER 2

Data Acquisition

We are held responsible for our actions, whether intentional or not.

Robert Fripp

2.1 INTRODUCTION

Before differentiating the considered methodologies, it is now absolutely fundamental to point out a basic and often little understood fact: the way dispersive properties are retrieved (which is the subject of this chapter) is something completely different and independent from the way these *data* are successively inverted.

It means that we can acquire data according, for instance, to the Multichannel Analysis of Surface Waves (MASW) approach but we will then have several ways to treat and invert the obtained velocity spectra. It is utterly imperative to keep these two facts (the determination of the velocity spectra and the successive inversion) completely separate.

Once we determine the dispersive properties, we can in fact exploit the obtained data according to different procedures and inversion strategies (see Chapters 5 and 6 and all the case studies in the Appendix).

Data acquisition is the series of field operations that will eventually influence the type and the quality of our analyses and, therefore, the accuracy of the retrieved subsurface model.

Quality and quantity are the two terms to be properly understood and managed. Field operations (i.e., the type of data we are going to acquire) must be performed carefully and jointly considering the goal of the survey and the characteristics of the area to investigate. That means that, in some cases, the acquisition can be quick and inexpensive (light equipment and limited acquisition procedures, i.e., time), while in other cases things can be much more complex and require a heavier effort (multichannel and multicomponent acquisitions and so forth).

As always, there is no *right* or *wrong* in universal terms and the survey must be planned according to the goals and local stratigraphic conditions (that can make the overall analyses simpler or harder).

Only a careful survey planning accomplished by personnel highly skilled in all the aspects related to the surface-wave phenomenology (Chapter 3) and consequent data processing can ensure the success of the analyses.

It is clear that the objective to pursue is the clear identification of the *optimal* solution for a given survey, which depends on the goals and the specific site stratigraphy. In this sense by *optimal*, we mean all the minimum number of components *necessary* and *sufficient*

Surface Wave Analysis for Near Surface Applications
ISBN 978-0-12-800770-9, http://dx.doi.org/10.1016/B978-0-12-800770-9.00002-9
23

to proceed with sound analyses not affected by ambiguities and capable of identifying a sufficiently constrained subsurface model.

As we try to underline throughout the whole book, there is no universal solution and the only way to proceed is by carefully considering all the theoretical aspects involved in seismic-data acquisition and processing, always keeping in mind the well-known *garbage in, garbage out* (GIGO) warning.

The problem here is to understand that *garbage* does not mean *noise* and, actually, *noise* does not mean anything (at least not in a universal sense).

To start understanding this point, Figure 2.1 presents a dataset before and after a simple low-pass filter is applied. It is quite apparent that, since surface waves used in geological/geotechnical applications are usually focused in the 4—50 Hz frequency range, the removal of the high-frequency "noise" has no significant role in surface-wave analysis and we could therefore ignore it (that is why we used the quotation marks).

So what can be a useful definition of noise that can help us in our analyses?

A very general and useful definition of noise can read as follows: noise is everything that is in our data while we would like not be there.

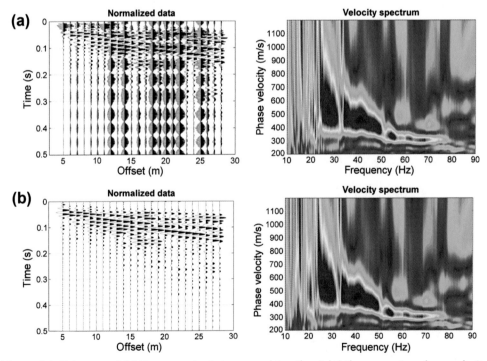

Figure 2.1 Noisy traces? (a) The raw seismic traces and (on the right) the respective phase-velocity spectrum. (b) The same dataset after a simple high-cut filter is applied. Traces are now much clearer but the velocity spectrum remains absolutely identical. Was the high-frequency "noise" a problem?

If we are dealing with reflection or reflection studies, surface waves are *noise*. If we are dealing with surface-wave analysis, reflections and refractions are *noise*. And so on.

A classical classification differentiates between *coherent* and *incoherent* noise. According to Telford et al. (1990), "coherent noise can be followed across at least few traces, unlike incoherent noise where we cannot predict what a trace will be like from a knowledge of nearby traces."

In general, if we perform the same acquisition (same source and same geometry) two or more times, incoherent noise will be different for different shots, while coherent noise tends to repeat the same way.

Finally, before starting to describe the active and passive methodologies commonly adopted for surface-wave acquisitions, a final point also requires some words.

In the so-called MASW, *Refraction Microtremor* (ReMi), and *Extended Spatial Autocorrelation* (ESAC) methodologies (which are based on multichannel acquisitions), we deal with *phase* velocities.

On the other hand, while dealing with Multiple Filter Analysis (MFA) or Frequency Time Analysis (FTAN) methodologies (typically—but not necessarily—based on the analysis of just one trace), we will deal with *group* velocities (i.e., the velocities associated not to a single frequency but to a more or less small package of frequencies). To understand the difference between phase and group velocities, see, for instance, Sachse and Pao (1978) or any introductory physics textbook.

2.2 ACTIVE METHODOLOGIES

In general terms, data acquisition for MASW is not too different than the common procedures adopted in refraction studies. It is actually simpler because, for instance, only *off-end* shooting is required (Figure 2.2).

Two trivial aspects to consider relate the possible trace saturation (clipping) and the attenuation of incoherent noise by means of a *vertical stack* (i.e., the summation of traces from multiple shots).

Figure 2.3 describes a well-known issue to face while using old non high-quality seismographs that, because of a limited dynamic range, apply a gain (i.e., a sort of amplification that can be modified by the user) to the signals coming from the geophones. In refraction studies (where the quantity that must be precisely identified is the arrival time), the saturation of the traces is not a problem and the data since the precise amplitude of the signals (which here appear truncated) is usually not relevant.

But dealing with surface-wave analysis (which, to some degree and during certain passages, requires the evaluation of the amplitude spectra), this can become a problem. As a consequence, while acquiring data for dispersion or attenuation analyses, saturation must be carefully avoided by fixing a proper amplification value (ideally the amplification

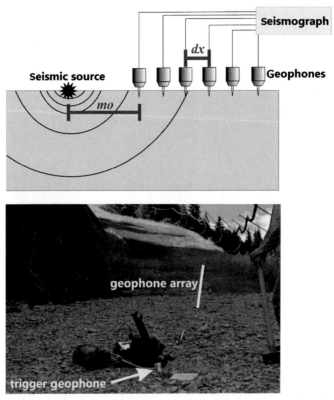

Figure 2.2 Classical geophone array (off-end shooting) used for body-wave refraction studies, as well as for surface-wave acquisition: *mo* indicates the *minimum offset* and *dx*, the *geophone spacing*. Different sources and receivers allow the acquisition of different components.

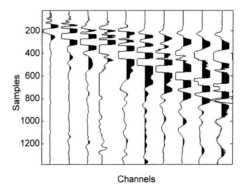

Figure 2.3 Clipped data: the amplitude exceeds the dynamic range of the seismograph. In order to avoid this, the *gain* applied to the traces must be decreased (or a high-quality seismograph must be used).

should be the same for all the channels) or by using a high-quality seismograph with an elevated dynamic range.

As for any other active technique (e.g., refraction and reflection studies), to attenuate the incoherent noise, *vertical stack* is always recommended (Figure 2.4).

Naming the data files during the acquisition (so already on the field) is a seemingly trivial thing that actually, if done in an unthoughtful manner, can result in cumbersome analysis procedures. On the field, it is in fact common practice to give file names that do not hold any intrinsic information. File names related to data acquisitions performed, for instance, in Berlin are then often something like *Berlin-shot1.dat* or *Ber-1.segy*. Of course, there is nothing wrong about that but this kind of naming is quite meaningless and requires ancillary information regarding the geometry and the considered component(s).

Here we propose and suggest a sort of simple "coding system" that can actually result extremely useful to speed up field operations and avoid misunderstandings with the fellows in charge of the data analysis.

Since the whole book tries to put in evidence that multicomponent analysis (thus acquisition) is a key point to consider if the goal is to provide well-constrained and unambiguous subsurface models, the very first aspect to clarify is about an efficient system to codify the considered component(s). Table 2.1 summarizes the meaning of the coding system somehow adopted by Herrmann (2003) in his codes and that we use all over the book and strongly recommend to everybody in the daily practice. In short, it allows describing the combination of the source and receiver(s) used in a very synthetic way (see also Figure 2.5).

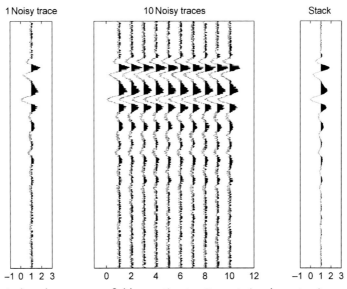

Figure 2.4 *Vertical stack*: a common field operation to attenuate incoherent noise.

Table 2.1 The Five Components Considered for Surface-Wave Analysis

Component	Geophone(s)	Source	Use
ZVF	Vertical (Z) (Figure 2.5(a))	Vertical Force (e.g., sledgehammer, Vibroseis, or weight drop)	Vertical component of Rayleigh waves
ZEX	Vertical (Z) (Figure 2.5(a))	EXplosive	Vertical component of Rayleigh waves
RVF	Radial (R)—axis of the horizontal geophone parallel to the array (Figure 2.5(b))	Vertical Force (e.g., sledgehammer, Vibroseis, or weight drop)	Radial component of Rayleigh waves
REX	Radial (R)—axis of the horizontal geophone parallel to the array (Figure 2.5(b))	EXplosive	Radial component of Rayleigh waves
THF	Transversal (T)—axis of the horizontal geophone perpendicular to the array (Figure 2.5(c))	Horizontal Force (shear source)	Love waves

First letter indicates the type and orientation of the geophones (Z, R, or T), while the second and third letters relate to the orientation and nature of the adopted seismic source, respectively (VF, EX, or HF).

The first letter refers to the considered geophone:

Z = vertical geophone

R = horizontal geophone set radially with respect to the source

T = horizontal geophone set transversally with respect to the array

While the second and third letters describe the source:

VF = vertical force (e.g., vertical-impact sledgehammer)

HF = horizontal force (e.g., horizontal-impact sledgehammer—see Figure 2.6)

EX = explosive source (rifle gun and so forth).

If we also want to provide information about the geometry (*minimum offset* and *geophone spacing*), we will end up with file names such as, for instance:

ZVF-mo4-dx3.dat

RVF-mo5-dx4.sg2

THF-mo5-dx4.segy

The first file name describes a common dataset often used in the so-called MASW analyses: vertical geophones and vertical-component source (e.g., sledgehammer) while considering a minimum offset of 4 m and geophone spacing of 3 m.

The second dataset was instead acquired to analyze the radial component of Rayleigh waves: horizontal geophones (set radially with respect to the array—see Figure 2.5(b)) were deployed each 4 m being the minimum offset of 5 m (again a vertical-component source).

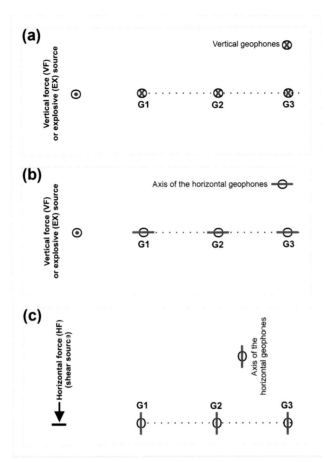

Figure 2.5 Sources and geophone orientation to adopt to obtain the five considered components (see Table 2.1): (a) vertical component of Rayleigh waves, (b) radial component of Rayleigh waves, (c) Love waves. Bird's-eye view.

Finally, *THF-mo5-dx4* simply means that the axis of the horizontal geophones is now set perpendicularly to the array (transversal component with the same geometry as before) and that we are now using a shear source (HF) as in Figure 2.6 (the different file extensions clearly refer to different seismic formats such as the *seg2* and the *segy*).

It is clear that the first and the second datasets are useful to analyze the vertical and radial components of Rayleigh waves, respectively and the third one relates to Love waves.

Almost useless to say that the second (RVF) and third (THF) datasets can be easily and quickly acquired by simply rotating the horizontal geophones and adopting the appropriate source (VF and HF) so that, by using horizontal geophones only, we can actually acquire both Rayleigh (their radial component) and Love waves.

Of course, by using 3-component (3C) geophones, we could acquire all the components in a quicker way, but if we are interested in a single vertical profile (seismic-hazard studies aimed at estimating the amplification factor for a single specific site typically require just a single 1D V_S vertical profile), that sort of purchase would be actually quite unnecessary.

Figure 2.6 Data acquisition for Love-wave and/or SH-wave (i.e., the horizontal-component of shear waves) refraction/reflection studies: cheap (but effective) shear source and (transversal) orientation of the horizontal-component geophones.

Please notice that by naming the field datasets according to this coding system, all the successive procedures will be much more straightforward and no misunderstanding will be possible while analyzing the data (data exchange between the staff involved in the acquisition and processing will be clearly easier because there will be no need of wordy accompanying notes).

2.2.1 Multichannel Acquisition (MASW)

The determination of surface-wave phase velocities requires the acquisition of multi-channel datasets and Table 2.2 summarizes the fundamental parameters to consider while dealing with active acquisitions. Figures 2.7 and 2.8 are used to comment about the recording time and the number of channel issues (which are both site-dependant quantities to fix also considering the goals of the survey).

Table 2.2 Acquisition Parameters for typical *Multichannel Analysis of Surface Waves (MASW)* Analyses

Minimum offset (distance between the source and first geophone)	5–20 m You may decide to acquire a couple of datasets by moving the source (you will choose the best dataset while analyzing the respective velocity spectra)
Geophone spacing	The point is the following: the length of the geophone array must be as long as possible. If the available space is, for instance, 75 m and you have 24 geophones, then you can fix the geophone distance as 3 m (with a minim offset distance of 5 m) Two critical facts to consider: 1. The maximum penetration depth can be roughly considered not larger than half of the geophone array 2. The number of geophones is not that relevant (see Figure 2.8 and related text)
Type of geophones	Vertical component of Rayleigh waves: vertical geophones (Figure 2.5(a)) Radial component of Rayleigh waves: horizontal geophones set radially to the array (Figure 2.5(b)) Love waves: horizontal geophones set perpendicular to the array (Figure 2.5(c)) *Eigenfrequency*: 4.5 Hz (or less)
Recording time	2 s are usually more than sufficient (it is essential that the full surface-wave trend is entirely recorded even at the very last channel/trace)—see Figure 2.7 and related text
Number of channels/geophones	12–24 Less number of channels are sometimes sufficient (see Figure 2.8 and related text) The crucial point is anyway the total length of the array, possibly not less than say 50 m, much better 70–90 m (to reach the suggested length, just act by modifying the geophone distance)
dt (sampling interval)	For near-surface geological applications, 1 ms is definitely more than sufficient (Nyquist frequency = 500 Hz)
Notes	No AGC (*Automatic Gain Control*) No filter Keep, if possible, the same amplification/gain value for all the channels (this becomes imperative when we are interested in attenuation analyses). Just be careful: do not saturate the channels close to the source (Figure 2.3) and keep a good signal-to-noise ratio for the distant offsets (where the amplitude of the signal is necessarily lower)

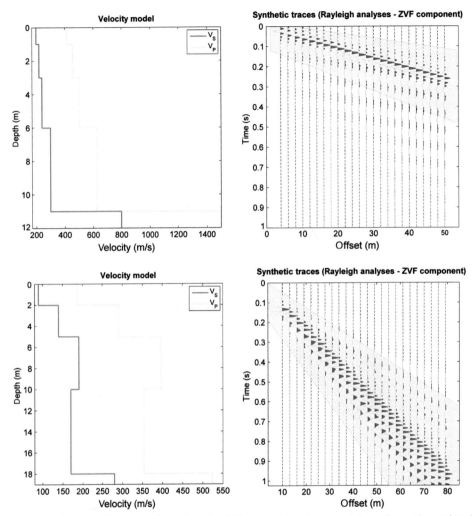

Figure 2.7 The recording time issue (on the left two subsurface models and, on the right, the computed synthetic traces for different acquisition geometry): not-too-soft sediments and/or short arrays require limited recording time (upper panel) (here 1 s is by far more than sufficient); very-soft sediments (i.e., very low shear-wave velocities and consequently surface-wave velocities) and/or longer arrays require longer recording time (here 1 s is not enough because at the distant offsets, surface waves arrive well after 1 s).

There is in fact a myth to dispel about the (minimum) number of geophones to use while acquiring surface waves for MASW (*lato sensu*) analyses. Very often it is believed that the more the better but while this idea can be somehow relevant for high-resolution reflection/refraction studies, the same does not hold for surface waves.

Figure 2.8 reports the seismic traces and the phase velocity spectra computed according to the phase-shift approach (Park et al., 1998; Dal Moro et al., 2003) while

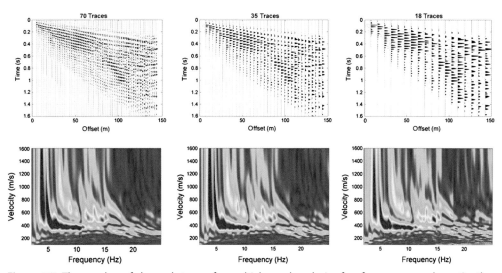

Figure 2.8 The number of channels to use for multichannel analysis of surface waves analyses. On the left column, the original 70-trace dataset and the corresponding phase-velocity spectrum; in the middle, the data after removing half of the original traces; on the right, the data after further removing half of the traces. Is there any significant difference in the velocity spectra?

considering 70, 35, and 18 traces, respectively. It is quite apparent that the velocity spectra are absolutely comparable (if not identical) and the very minor differences are simply related to the fact that, dealing with a smaller number of traces, the summation (i.e., averaging) process intrinsic in the velocity spectra computation via phase shift is slightly more influenced by "noise" related to scattered events and so forth (please note that the array considered in Figure 2.8 is quite long—almost 150 m—and that the very last traces show clear evidences of some local discontinuity).

The moral is quite straightforward: very often, if the length of the array is not too large, 12 traces are absolutely sufficient and there is no need for a larger amount of channels that are instead important if the dispersive properties are determined via $f-k$ transform, which is clearly subject to problems related to *spatial aliasing* (see Dal Moro et al., 2003).

Further examples are reported in Chapter 3 (while speaking about the *spatial aliasing*) and in Chapter 6 (while speaking about the *Full Velocity Spectra* inversion).

It must be always considered that the total length of the array fixes the maximum penetration depth, which can be roughly estimated in half of the length of the array (which is the distance between the first and the last geophones).

2.2.2 Single-Channel Acquisitions (MFA/FTAN)

If the goal is the determination (and successive analysis) of the group-velocity spectrum, only one trace is actually necessary (Dziewonsky et al., 1969; Natale et al., 2004). This means that a common 3C geophone for *Horizontal-to-Vertical Spectral*

Figure 2.9 Data acquisition for the retrieving group velocities via frequency–time analyses.

Ratio (HVSR) measurements (Chapter 4) can be proficiently used (if equipped with a triggerable system that fixes the zero time) also for retrieving the group velocities (see Chapter 3). Estimating the maximum penetration depth for this kind of analyses is, as for any other, quite hard cause a number of aspects actually come into play. Experience suggests that a reasonable value that can be adopted as the umpteenth rule of thumb is that, if we do everything right and can also use the RVSR curve, it is possible to hope to determine the V_S profile down to a depth of about two thirds of the adopted *offset* (see for instance case study 2 and 6), while about half of it in case we can use only the dispersion data, without the RVSR curve (e.g., case study 8).

Figure 2.9 highlights the radial and transversal components (the vertical one does not need any explanation) in case a 3C geophone is used for this kind of active acquisitions (compare with schemes reported in Table 2.1 and Figures 2.5 and 2.6).

Actually, if we are dealing with a 3C geophone, we can eventually consider four components. Figure 2.10 reports the "objects" that we can determine in this case

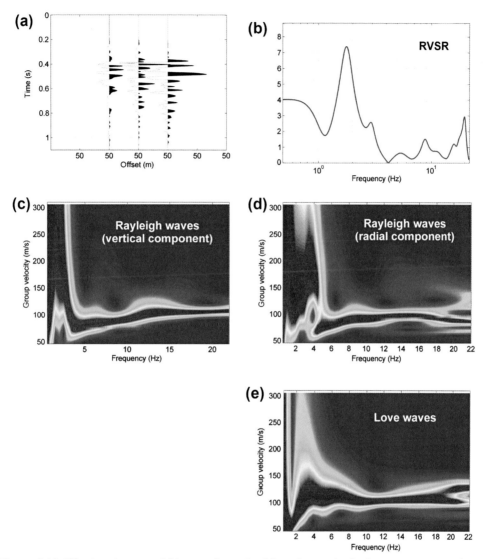

Figure 2.10 Effective data acquisition performed while using a single 3-component geophone (calibrated and triggerable). Acquired data can be used for the holistic analysis of surface-wave dispersion via group-velocity spectra analysis: (a) the three seismic traces (vertical, radial, and transversal components); (b) *Radial-to-Vertical Spectral Ratio* (RVSR); (c) and (d) the group-velocity spectra for the vertical and radial components of Rayleigh waves; (e) THF component (Love waves). Their joint analysis is a patent-pending methodology (Dal Moro, 2014b). See Paragraph 7.2.2 and case studies 2 and 8.

(and eventually consider in our joint inversion procedures—see Chapters 5 and 6): the radial (RVF or REX) and vertical (ZVF or ZEX) components of Rayleigh waves, their *Radial-to-Vertical Spectral Ratio* (RVSR) and Love waves (THF).

2.3 PASSIVE METHODOLOGIES

In the low-frequency range, multichannel active acquisitions often lack clear information, and a series of passive techniques aimed at identifying the velocities in the low-frequency range was then implemented in the last decade. A number of techniques are surely available, but here we will just clarify a couple of basic points related to two of the most common ones.

It goes without saying that, since the focus is on the low-frequency range, it is mandatory to adopt good sensors (geophones) characterized by low *eigenfrequency*: 4.5 Hz is the very minimum value but the survey would clearly benefit from the use of lower-frequency sensors (actually quite expensive).

Please note that a geophone is anyway capable of retrieving signals even at frequencies lower than its *eigenfrequency* and an often-used rule of thumb proposes an actual "visibility limit" half the formal *eigenfrequency*.

The most popular passive technique based on linear arrays is probably the ReMi (*Refraction Microtremor*—Louie, 2001). Because of its practical implementation (a linear array) and considered that the locations of the sources are necessarily unknown (passive technique), the consequence is that the obtained velocity spectra must be interpreted following a criterion of the "lower velocity bound." In fact, considered a linear array (see Figure 2.11), the *B* source (considered at a very large distance from the array) will

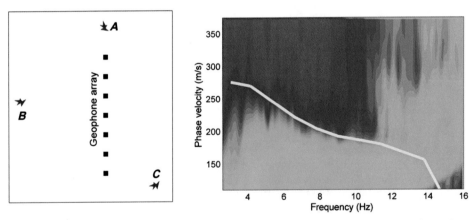

Figure 2.11 ReMi acquisition and analysis: the linear geophone array and the lower-velocity bound criterion for the interpretation of the obtained velocity spectrum. The yellow line reported over the ReMi velocity spectrum is the apparent dispersion curve identified through the Extended Spatial Auto-correlation analyses reported in Figure 2.12(c) (clearly for the same site). The color scale of the ReMi velocity spectrum was chosen to emphasize the lower-velocity border. Main acquisition parameters for passive acquisitions aimed at depicting surface-wave dispersion are summarized in Table 2.3.

Table 2.3 Main Acquisition Parameters for ReMi and ESAC (and Related) Techniques

Geometry	In general terms, the larger the better but see text for details
Type of geophones	For Rayleigh waves: vertical geophones
	For Love waves: 3-component sensors (see Köhler et al., 2007; Poggi and Fäh, 2010)
	Eigenfrequency: 4.5 Hz or (better) less
Record time	10–30 min
Number of channels/geophones	16–24
dt (sampling interval)	2–6 ms
Notes	As usual no AGC, no filter
	For some recommendations about the gain see text

AGC, automatic gain control; ReMi, Refraction Microtremor; ESAC, Extended Spatial Autocorrelation.

clearly produce an infinite apparent velocity (all the signals will reach the geophones at the same time). On the other side, the *A* source (again to be considered at an infinite distance) will produce the smallest velocities that represent the "lower bound" of the signal in the ReMi velocity spectrum. The problem is that such a "lower bound" is clearly necessarily fuzzy and ambiguous and to some degree also influenced by the color scale used while representing the ReMi velocity spectrum.

On the other side, ESAC technique relies on bidimensional arrays of arbitrary shape and, as the acronym suggests, represents an extension (i.e., a generalization) of a technique based on circular/symmetric arrays (Spatial Autocorrelation; e.g., Asten, 1978; Asten and Henstridge, 1984; Okada, 2003, 2006). Very briefly, the ESAC approach allows determining phase velocities through the evaluation of the Bessel functions for each considered frequency (for details see Ohori et al., 2002).

For ESAC acquisitions, sensors can be deployed in a completely random fashion (for instance, following logistic aspects related to complex urban environments where simple geometries could be not viable) but, when possible, the simplest way to acquire data useful for ESAC analyses is an L-shaped array like the one reported in Figure 2.12(a).

This sort of geometry is in fact extremely simple to execute on the field and, furthermore, by placing a seismic source in the *[0, 0]* point (red star-like point in Figure 2.12(a)), we can also acquire data useful for two active (MASW) analyses along the two perpendicular, A and B, branches.

Figure 2.12 also reports the distances between pairs of channels considered for computing the correlation factors to compare with the Bessel functions (the different distances ensure that all the wavelengths are sufficiently covered) and the final outcome (a velocity spectrum with its *apparent* dispersion curve—please note that, differently than with the ReMi approach, now the apparent dispersion curve is represented by the peak values in the velocity spectrum and these are then not subjected to any personal evaluation).

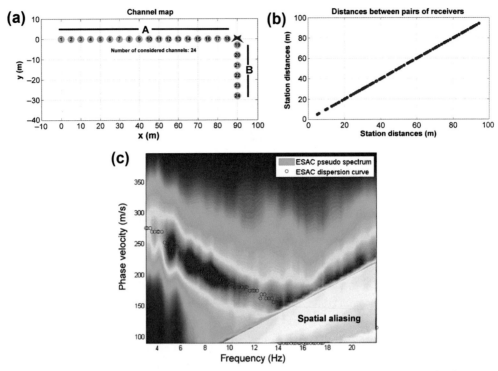

Figure 2.12 Extended Spatial Autocorrelation (ESAC) analysis: (a) bidimensional array, (b) distances between considered pairs of channels, (c) retrieved apparent dispersion curve (also indicated the *spatial aliasing* area).

It is quite important to underline the fact that the fundamental difference with respect to the active techniques is that now the sources are not only unknown but anyway more distant with respect to common active acquisition. The offsets are thus much larger and this can mirror in *apparent* dispersion curves where the excited (most energetic) modes can be slightly different than for the active acquisitions.

Anyway, the *apparent* dispersion curves retrieved from passive methodologies is in general result of multiple modes (see Chapter 3 and several of the case studies in the Appendix) and under no circumstances can be considered as necessarily representative of the fundamental one (see, e.g., case study 12 and Dal Moro, 2011).

Here, as well as for the active techniques introduced in the previous sections, the penetration depth is clearly proportional to the array dimension (which is the distance between the two most distant geophones) and can be roughly estimated (once again a simple but useful rule of thumb) in approximately half of the array length (or slightly more).

While acquiring datasets according to passive techniques, some care must be paid to the gain applied to the traces on the field during the acquisition procedures (see the previously described problem about data truncation—Figure 2.3). Usually, the signals forming the background microtremor field that we record to determine surface-wave phase velocities have traveled a long way before reaching our sensors and, consequently, their amplitudes have undergone a relevant attenuation which simply mean that they are usually quite small. If we are using an old-style seismograph with a limited dynamic range (with modifiable gain values), it is recommendable to fix a high amplification value that will give the chance to highlight the low-amplitude signals of the microtremor background (also clearly avoiding trace saturation). Of course, thanks to the very wide dynamic range, up-to-date 24-bit seismographs have solved these problems and do not require any maneuver from the user.

Because of attenuation of the high-frequency components and *spatial aliasing*, passive techniques are typically unable to retrieve the dispersion curves in the high-frequency range. Consequently, the sampling rate can be reduced to smaller values: 166 Hz (sampling interval about 6 ms) is therefore usually more than sufficient. On the other side, the recording time must be clearly much larger than for the active case and 10–30 min are usually recommended.

Finally, it should be pointed out that, by using vertical-component sensors, it is clearly possible to retrieve the phase velocities of Rayleigh wave only. To properly detect Love waves, it is necessary to use 3C sensors whose data, properly processed (e.g., Köhler et al., 2007; Poggi and Fäh, 2010), can identify Love-wave dispersion (see the case study 6).

2.3.1 Acquiring Data for the Determination of the H/V Spectral Ratio

Although the acquisition procedures aimed at determining the *HVSR* (see Chapter 4) are quite simple, nevertheless they must be carefully accomplished. First of all, in order to acquire useful data, the 3C sensor to use must be properly calibrated and tested (Tasic and Runovc, 2010) and the field procedures to follow can then be summarized in a few schematic points:

1. If possible, avoid working on asphalt (or, in general, on any superficial stiff layer) because that would somehow alter the HVSR (see Chapter 4).
2. In order to remove very shallow stratigraphic accidents, it can be useful to dig a small hole and place the sensor inside (Figure 2.13(a)).
3. Carefully check the bubble level (the geophone must be perfectly horizontal) (Figure 2.13(c and d)).
4. In case of nonnegligible wind, protect the equipment from it (Figure 2.13(b)).

When possible, avoid staying too close to natural or artificial structures such as big trees, large and tall buildings, and so forth.

Figure 2.13 Careful data acquisition for horizontal-to-vertical spectral ratio analyses. (a) Dig a small hole and place the 3-component geophone inside, (b) protect the geophone from wind, (c) ensure that the geophone spirit level bubble is centered, (d) geophone is not properly leveled (bubble is not centered).

Figure 2.14 Cheap solutions for data acquisition on asphalt cover (for long 2D acquisitions land-streamers are clearly a must - O'Neill et al., 2006).

2.4 FEW FINAL REMARKS

We should never forget that the quality of the subsurface model eventually identified depends on several elements: a careful survey planning, the proper accomplishment of all the field procedures and, eventually, the way analyses are performed.

The final model (its accuracy) actually depends on the weak ring of a long sequence of choices and operations:

1. A very long array (which is in general important to pick up low frequencies) can be useless if, during an active acquisition, the adopted source does not produce sufficient low frequencies (that is why for acquiring low frequencies, passive methods are often preferred).
2. The acquisition of one component only typically prevents from properly constrained inversions.
3. Using a large number of geophones is rarely important (see Figure 2.8, Chapter 6 and Dal Moro et al., 2003).
4. Dispersion analyses can be performed by following different approaches that, in order to choose the most appropriate, we should master (the whole book intends to give a scenario of several possible inversion strategies).

Of course, the very final possible (and most dangerous) weak ring is represented by the *know-how* of the person involved in the data analyses because the most perfect and complete datasets are useless in the hands of the careless incompetent.

While planning an acquisition campaign, we should always remember that the goal is to provide the processing center with good data both in terms of quality and quantity and, since forgetting something essential while being tens or thousands of kilometers from the office is quite risky, a possible (surely incomplete) *checklist* to personalize according to specific needs is provided.

Seismograph		✔
Seismograph battery(ies)		✔
Trigger geophone and trigger cable		✔
Seismic cable(s)		✔
Depending on the component(s), we decided to acquire	Vertical geophones	✔
	Horizontal geophones	✔
	3-Component geophone(s)	✔
Accessories for acquisitions on asphalt (Figure 2.14)		✔
Plate (for VF acquisitions) and/or beam (for HF acquisitions)		✔
Sledgehammer or rifle gun		✔
3-Component geophone for HVSR measurements (+accessories)		✔
Notebook		✔
Pens/pencils		✔
Photo camera		✔
Miscellanea: umbrella, GPS, foldable seat (fisherman's type), food and beverage, toolbox (screw drivers, hammers, Swiss army knife, etc.), and so forth		✔

HVSR, horizontal-to-vertical spectral ratio; GPS, global positioning system.

CHAPTER 3

Understanding Surface-Wave Phenomenology

If you don't know, why do you ask?

David Tudor

3.1 INTRODUCING THE PROBLEM

This chapter probably represents the very core of the entire book which, in our intentions, is meant to fill the gap between formal academic research and practical near-surface applications.

As declared in the introduction, the decision not to report any equation was taken for two reasons:

1. All the math can be easily found in a number of papers (see mentioned references).
2. The main goal is to provide a set of evidences of nontrivial datasets whose complexity can be clearly explained in terms of physical/mathematical facts but that we are here interested in understanding in the practical consequences.

In fact, by using the term *phenomenology*, we intend to underline the idea that the way the physical phenomena must be read depends on a conscious effort that is necessarily based on a solid theoretical background.

The way surface waves propagate and disperse (and attenuate) can be in fact extremely complex and any simplistic assumption based on a limited theoretical basis and/or field experience, inevitably leads to meaningless analyses.

In this book, we decided to consider both surface-wave dispersion (according to different acquisition and inversion procedures) and *horizontal-to-vertical spectral ratio* (HVSR). The reason is twofold: on one (practical) side, both these methodologies are more and more used for a number of geotechnical applications and, on the other (theoretical) side, surface-wave dispersion and HVSR are two aspects of the same phenomenon since, in both cases, we are considering surface-wave propagation.

On one side, dispersion focuses on the propagation velocity and depends on the local velocity structure and on the other side, HVSR is related to the relative amplitude of surface waves present in the background microtremor field and depends on both the local subsurface velocity structure and on the dominating source characteristics of the microtremor field responsible for the observed HVSR.

To some degree, this latter is therefore somehow time-dependent and different meteorological/seasonal conditions can reflect in different HVSR curves (see Chapter 4) but,

Surface Wave Analysis for Near Surface Applications
ISBN 978-0-12-800770-9, http://dx.doi.org/10.1016/B978-0-12-800770-9.00003-0
43

otherwise, surface-wave dispersion and HVSR can be regarded as two aspects of the same phenomenon: in one case, we concentrate on the velocities and in the other, on the amplitudes (that, being the source locations and spectral signatures unknown, can be considered only in a relative sense—the amplitude of the horizontal components with respect to the vertical one).

The point that represents the very crucial aspect while analyzing surface waves is represented by the presence of different modes of propagation that, if not properly considered, inevitably lead to erroneous reconstructions of the subsurface conditions (e.g., Zhang and Chan, 2003). Few synthetic datasets can be profitably used to get familiar with these fundamental issues.

In Figure 3.1, a first synthetic dataset reports a case where the sequence (from low to high frequencies) of modes is the following: up to 35 Hz, the dominating mode is the fundamental one (mode0), then (up to 52 Hz) dominates the second higher mode (mode2), and then (for frequencies higher than about 52 Hz) the dominating mode is the first overtone (mode1). Two mode jumps are then present: the first from mode0 to mode2 (at about 35 Hz) and the second from mode2 to mode1 (at about 52 Hz).

It is apparent that the energy can unfold while jumping in a so-to-speak nonsequential way: the mode jump does not have to be necessarily from mode0 to mode1 or from mode1 to mode2 (or vice versa), but specific stratigraphic conditions can produce a jump from the mode n to the mode $n + m$ (being m higher than 1) and there is no reason to assume that if a mode jump occurs, this is necessarily from the mode n to the mode $n + 1$.

Actually, since the apparent dispersion curve depends also on the distance between the source and the receiver(s) (e.g., Tokimatsu et al., 1992), at the same site, a different acquisition geometry can produce a different velocity spectrum.

Let us now consider the dataset reported in Figure 3.2. Here the signal dominating the velocity spectrum for frequencies higher than 20 Hz is clearly continuous. Nevertheless, if we consider the overlaying modal dispersion curves, we can easily notice that, as a matter of facts, such a signal is the result of the superposition of two modes (the first higher mode for frequencies lower than 40 Hz and the fundamental one for higher frequencies).

This gives us the evidence that continuity of a signal does not necessarily mean uniqueness of the responsible mode.

Often, surface-wave dispersion is analyzed through the *picking* of one or more dispersion curves that a practitioner defines giving his/her own interpretation in terms of modes (i.e., modal dispersion curves). Often, it is picked a fundamental mode and, if present, some higher mode. But in both the presented datasets (Figures 3.1 and 3.2), this operation would be actually quite problematic. In particular, in the second case, such a common (and simplistic) procedure (picking of the modal dispersion curves) would clearly be quite dangerous since erroneous interpretations would be practically inevitable.

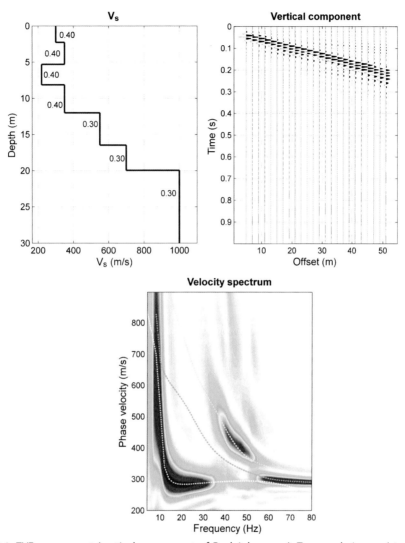

Figure 3.1 ZVF component (vertical component of Rayleigh waves). Two mode jumps (at about 35 and 52 Hz) are apparent: the first from the fundamental to the second higher mode, the second from the second to the first higher mode. Synthetic traces computed according to Carcione (1992). Overlaying modal dispersion curves are computed according to Dunkin (1965).

Data reported in Figure 3.3 are presented to illustrate both a further case of continuity not associated to a single mode both the way *spatial aliasing* shows up: in the upper panel (while considering 48 traces), the signal in the velocity spectrum is continuous but actually related to two different modes; in the lower panel (while considering 24 traces), the "real signal" does not modify (even if continuous and without any apparent "mode jump", the *effective* dispersion curve is still related to 2 different modes) but an artifact

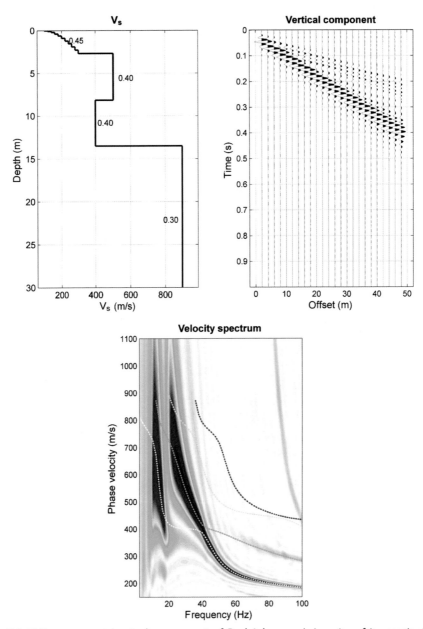

Figure 3.2 ZVF component (vertical component of Rayleigh waves). In spite of its continuity, the signal between 20 and 100 Hz is actually the result of two modes (the fundamental for frequencies higher than 40 Hz, the first higher mode for frequencies between 20 and 40 Hz). Synthetic traces computed according to Carcione (1992). Overlaying modal dispersion curves are computed according to Dunkin (1965).

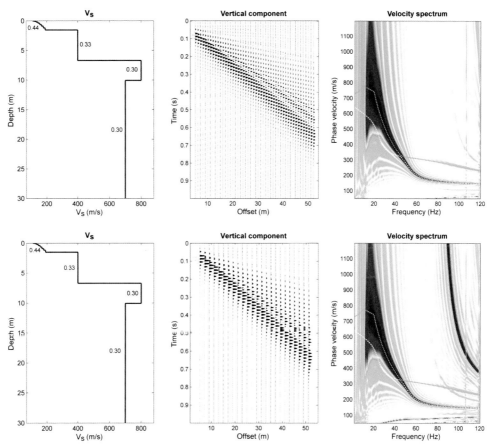

Figure 3.3 ZVF component (synthetic traces computed according to Carcione (1992)) for the same model but considering a different number of traces (the total length of the array is the same): 48 traces in the upper panel and 24 traces in the lower panel. In the latter case, *spatial aliasing* takes place: the "real" part of the velocity spectrum is unaltered (which is the relevant point) but a spurious signal shows up in the very high frequency–velocity part of the velocity spectrum (since this latter signal is very easily identified, its presence is not dangerous or harmful). Overlaying modal dispersion curves are computed according to Dunkin (1965). Once again, in spite of its continuity and the lack of any apparent "mode jump", the signal in the velocity spectrum pertains to two different modes.

(*spatial aliasing*) appears. Its distinctive characteristics (very defined narrow signal at very high frequencies and velocities) do not anyway represent a problem since its identification as artifact is definitely straightforward.

For the data reported in Figure 3.4, we considered a superficial stiff later over a softer stratum characterized by very high Poisson's ratio (we could, for instance, imagine a landfill layer over a saturated sand or peat): the aspect of the obtained velocity spectrum is quite specific (moving from the low to the high frequencies, there are portions of higher

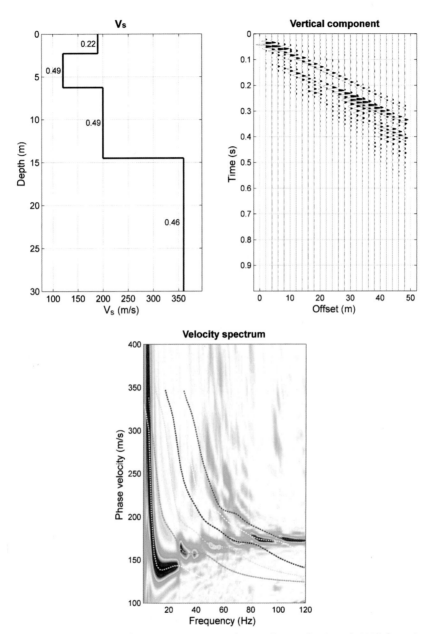

Figure 3.4 ZVF component (synthetic traces computed according to Carcione (1992)) for a site where a shallow V_S inversion occurs and the underlying low-velocity layer is characterized by large Poisson's moduli. The characteristic mode excitement is often referred to as *mode splitting*. Overlaying modal dispersion curves are computed according to Dunkin (1965).

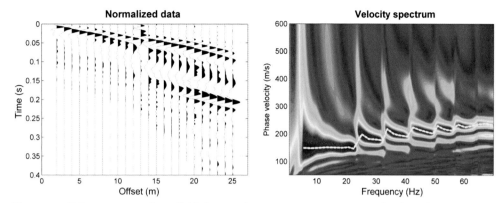

Figure 3.5 ZVF component for a field dataset showing an apparent example of *mode splitting*. The yellow dotted line represents the so-called *effective* (or *apparent*) dispersion curve which can be computed (thus inverted) considering the mathematical formulation of Tokimatsu et al. (1992), without any interpretation in terms of modal curves.

and higher modes) and, when such a trend is particularly manifested, is sometimes referred to as *mode splitting* (see also O'Neill and Matsuoka, 2005). A field dataset showing this sort of trend is reported in Figure 3.5. Please notice that the similarity is apparent not only in the velocity spectrum but on the seismic traces themselves (offset—time domain): in both cases, there is a sort of "double wavelet".

Incidentally, from a computational point of view, a key quantity responsible for the actual distribution of energy among the modes is represented by the I1 energy integral (e.g., Snieder, 2002), but the offsets (and, more precisely, the geophone spacing) are also relevant, being the *apparent* (or *effective*) dispersion curve a function also of that (e.g., Tokimatsu et al., 1992).

Speaking about Rayleigh waves, it is imperative to clearly point out that the apparent dispersion curve is different for the radial and vertical components (see again Tokimatsu et al., 1992). The difference can be minimal or extremely large, depending on the local stratigraphic conditions (see next Paragraph). From the practical point of view, this will be clear by comparing, for instance, the case study 6 (RVF and ZVF are significantly different) and the case study 2 (the differences between RVF and ZVF are minor).

3.2 MORE ABOUT MODES AND COMPONENTS

3.2.1 Phase Velocities

The datasets seen so far are those of the most common component used in the near-surface applications: ZVF, i.e., the vertical component of Rayleigh waves obtained while using a vertical-impact force and vertical-component geophones.

Anyway, since Rayleigh waves have also a radial component, it is now time to open up a bit more our vision on surface-wave propagation and dispersion. Figure 3.6 reports

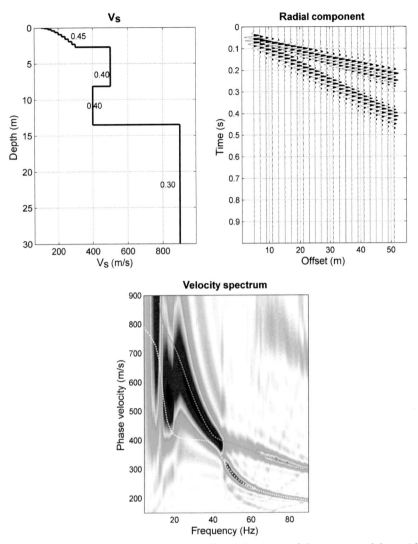

Figure 3.6 RVF component (radial component of Rayleigh waves) of the same model considered in Figure 3.2. Energy distribution among the modes (i.e., the mode jumps) is different for the two (vertical and radial) components.

the radial component for the same model considered for the data presented in Figure 3.2 (where it was considered the vertical component). As already pointed out, the velocity spectra (i.e., the apparent dispersion curves) of the radial and vertical components of Rayleigh waves are in general different and this will result crucial for properly planning our acquisitions and analyses, but some more aspects surely need to be clarified.

So far we have seen datasets where low-velocity channels or stiff layers were present, and we might risk inferring that complex velocity spectra are the result of this sort of stratigraphic features.

In the next pages, in order to disprove this simplistic assumption and point out the effect of attenuation (Panza, 1989; Lai and Rix, 1998; Tonn, 1991; Barton, 2006), we will compute some synthetics according to the modal summation approach followed by Herrmann (2003) that, although approximate (compare for instance Lai and Rix, 2002), it results definitely sufficient for putting in evidence few important aspects (see also e.g., Malagnini et al., 1994, 1997).

Figure 3.7 reports the first velocity and quality factor model we will consider and can be thought of as a case where a soft sediment cover (about 4 m thick) lies over a weathered rock formation ($V_S = 600$ m/s) that becomes more rigid and consistent at a depth of about 14 m ($V_S = 1200$ m/s). No low-velocity layer (LVL) or stiff layer is present (acquisition geometry: 48 traces with geophone spacing of 1 m and minimum offset of 5 m).

For this model, we computed the synthetic traces and the respective phase-velocity spectra for the vertical and radial components of Rayleigh waves (ZVF and RVF) and Love waves (THF) (Figure 3.8). It is quite remarkable how, in spite of the fact that no LVL or stiff layer is present, the ZVF component is actually quite complex: the apparent dispersion curve is basically continuous, but is actually the result of two different modes: the fundamental (for frequencies higher than 20 Hz and lower than 10 Hz) and the first higher modes (between 10 and 20 Hz). On the other side, the RVF component is largely dominated by the fundamental mode only (a weak evidence of the first higher mode is present for frequencies higher than about 25 Hz) as well as Love waves (THF).

It is vital to point out that while in this case, the radial component of Rayleigh waves results "better" when compared to the vertical one (in the sense that its interpretation is easier in terms of modal curves), this evidence cannot, under any circumstance, be generalized. Depending on the specific stratigraphic conditions, ZVF can be "better" or "worse" than RVF or, in some cases, equivalent.

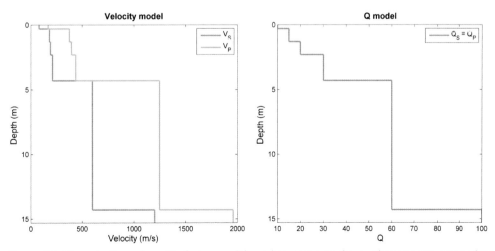

Figure 3.7 The velocity and quality factor model used to compute the synthetic traces reported in Figure 3.8. Please note that, no velocity inversion or stiff layer is present.

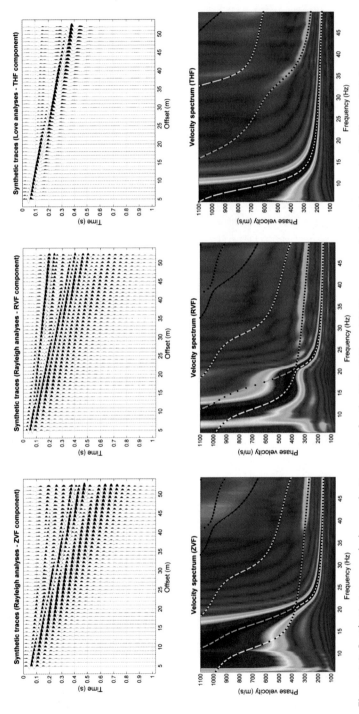

Figure 3.8 Synthetic traces and phase-velocity spectra for the ZVF, RVF, and THF components (model reported in Figure 3.7): Love waves are dominated by the fundamental mode (the first higher mode barely visible in the high-frequency range); radial and vertical components show a different distribution of energy among the different modes. In this case, while the radial component (RVF) is dominated by the fundamental mode, the vertical one (ZVF) is actually a complex mixture of the fundamental and first higher mode.

All those who are familiar with Love waves both from the theoretical and practical points of view are quite aware that, compared with Rayleigh waves, Love waves are invariably so-to-speak simpler to understand. Differently than Rayleigh waves, the fundamental mode is practically ubiquitous and the interpretation is then quite easy (see also case studies 4 and 11 and Safani et al., 2005).

A second and final model with increasing velocities (no LVLs or stiff layers) is now considered (Figure 3.9) with the same acquisition geometry as for the previous case.

Figures 3.10 and 3.11 now show the ZVF, RVF, ZEX, REX, and THF components: synthetic traces, corresponding velocity spectrum, and, overlaying, the modal dispersion curves for the first six modes (the explosive source is considered at a depth of 0.2 m). The most striking evidence is that the ZVF velocity spectrum (that means the vertical component of Rayleigh waves, nowadays the most common kind of multichannel analysis of surface waves (MASW)) does not show any clear evidence of the fundamental mode. If we would then consider only such a component, the consequence would be quite serious: we would risk interpreting the first higher mode as the fundamental one, thus eventually obtaining a completely erroneous V_S profile (shear-wave velocities would be severely overestimated). This is actually a quite common mistake, and a lot of MASW surveys often overestimate shear-wave velocities just because of this sort of phenomena.

On the other hand, the RVF component shows clear mode jumps (at 25 and 48 Hz) that make the velocity spectrum complex but yet definitely easier to interpret when compared to the ZVF component.

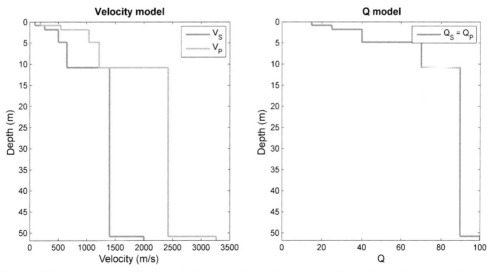

Figure 3.9 The velocity and quality factor model used to compute the synthetic traces reported in Figures 3.10 and 3.11. Please note that, again, no velocity inversion or stiff layer is present.

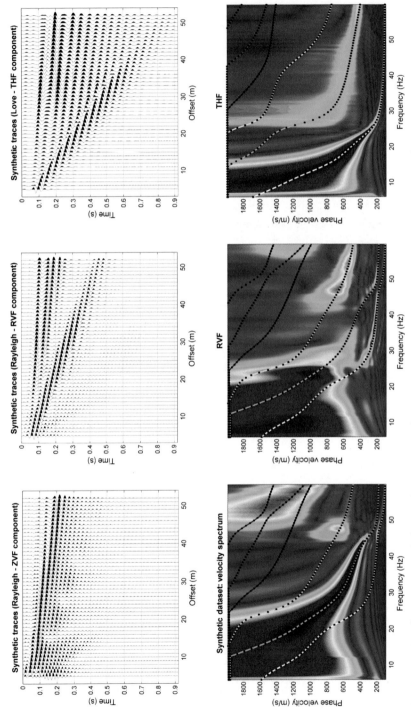

Figure 3.10 Synthetic traces and phase-velocity spectra for the ZVF, RVF, and THF components (model reported in Figure 3.9): Love waves are dominated by the fundamental mode (the first higher mode barely visible in the high-frequency range); radial and vertical components show a different distribution of energy among the different modes. In this case, while the radial component (RVF) is dominated by the fundamental mode, the vertical one (ZVF) is actually a complex mixture of fundamental and first higher mode. Shown the modal dispersion curves of the first six modes.

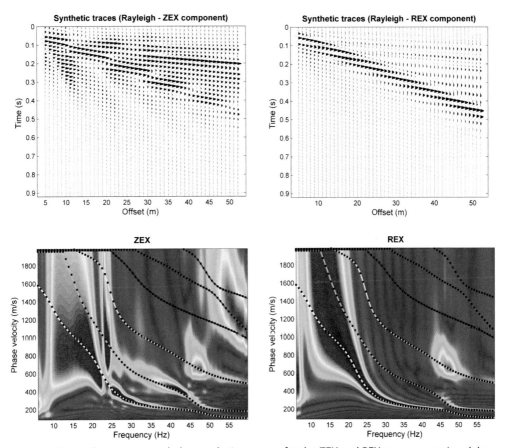

Figure 3.11 Synthetic traces and phase-velocity spectra for the ZEX and REX components (model reported in Figure 3.9): in this case, the explosive source determines velocity spectra that appear quite easy to understand in terms of modal curves.

As evident from the phase-velocity spectra reported in Figure 3.11, in this case, the use of an explosive source (not very common in surface-wave acquisition) would produce velocity spectra easily interpretable.

Incidentally, we should again underline the very simple behavior followed by Love waves compared to Rayleigh waves (see Figures 3.10 and 3.11), also recalling the fact that the joint acquisition of the RVF (radial component of Rayleigh waves) and THF (Love waves) can be inexpensively and quickly performed simply by rotating the horizontal geophones by 90° (we do not need vertical geophones to record Rayleigh waves, since we can record the radial component—see Chapter 1 and 2).

It would be clearly quite redundant to recall that the specific apparent dispersion curves are a function of the V_S profile (and of the acquisition geometry), but it is probably not completely useless to underline that also the attenuation plays a significant role (Panza,

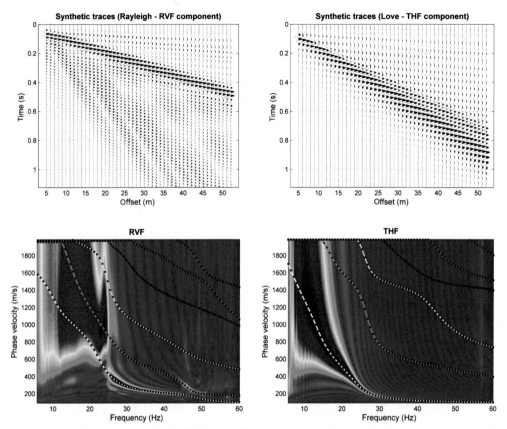

Figure 3.12 Velocity spectra for the RVF and THF components while considering the same V_S model reported in Figure 3.9 but infinite quality factors (elastic case). It is clear that higher modes are now less important when compared to the viscoelastic case considered in Figure 3.10. Shown the modal dispersion curves for the first six modes.

1989; 2011 personal communication). Figure 3.12 reports the RVF and THF phase-velocity spectra for the same V_S model (Figure 3.9) but now considering the purely elastic case (which practically means infinite quality factors). By comparing the data reported in Figure 3.12 with those shown in Figure 3.10, it is possible to note that now (elastic case) higher modes are less prominent. In other terms, the attenuation (which inevitably takes place in the real world) emphasizes higher modes (e.g., Panza, 1989) and, while analyzing surface-wave propagation, its role should then be properly considered.

3.2.2 Group Velocities

Most of the active and passive methodologies nowadays commonly used to retrieve the dispersive properties are based on trace-correlation techniques and rely on multichannel acquisitions. On the other side, seismologists have often to deal with signals (such as, for instance, those produced by *teleseismic* events) recorded by a single (typically 3-component) seismometer. For this reason, they also developed techniques capable of retrieving the dispersive properties while considering single traces. These techniques (e.g., Dziewonsky et al., 1969; Pedersen et al., 2003; Natale et al., 2004) are fundamentally based on data processing on the time—frequency domain and allow in determining the *group* velocities (not the *phase* velocities) as a function of the frequency.

Dealing with group-velocity spectra rather than with phase-velocity spectra is quite different (to get familiar with the differences between phase and group velocities, see any introductory physics textbook or, for instance, Sachse and Pao, 1978). While the reading (i.e., understanding) of the phase-velocity spectra is relatively easy (see Chapter 1), group velocities behave in a completely different way and the comprehension of the respective velocity spectra can be quite problematic when the way the energy jumps from one mode to the other is not trivial.

An example of simple dataset for which we computed both phase and group-velocity spectra is reported in Figure 3.13. Group-velocity spectrum is computed via *Multiple Filter Analysis* (MFA—Dziewonsky et al., 1969) while considering only the very last trace (offset 72 m).

The interesting aspect is clearly related to the fact that a single trace is necessary, while serious interpretative issues can appear when the way energy unfolds for different modes in different frequency ranges is complex.

Figure 3.14 reports, for instance, a synthetic case for which, in two definite frequency ranges, higher modes travel slower than low-order modes (see 16—21 Hz and 33—47 Hz frequency ranges) so that the interpretation of the group-velocity spectra in terms of modal dispersion curves can result quite tricky (please remember that, on the contrary, while considering phase velocities, higher modes are necessarily faster than low-order modes).

From the practical point of view, this means that, while considering group-velocity spectra, Love waves (which, as previously showed and underlined, typically propagate following a simpler behavior) should be considered as a mandatory component to acquire and that the group-velocity inversion would highly benefit from nonstandard approaches based on the effective dispersion curve or on the *Full Velocity Spectrum* approach (see Chapters 6 and 7 and case studies 2, 6, 8, and 12).

3.3 ABOUT PASSIVE METHODS

The velocity spectra or the apparent (or effective) dispersion curves that can be obtained while adopting some of the various passive methodologies basically undergo the same problems that we have seen for the active methods: the fundamental mode is not necessarily the dominating one and the continuity of a signal does not necessarily mean that a single mode is involved.

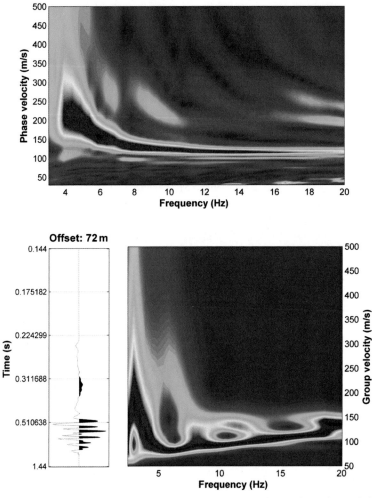

Figure 3.13 Comparison between the phase-velocity spectrum computed via phase shift (Park et al., 1998; Dal Moro et al., 2003) for a multichannel dataset (see case study 2) and the group-velocity spectrum computed via multiple filter analysis (Dziewonsky et al., 1969) by considering just the very last trace (offset 72 m).

The only difference between active and passive methods aimed at determining the phase velocities is in fact simply represented by the different offsets involved (in the passive methods, we are typically dealing with signals that, before reaching our geophone array, have traveled hundreds or thousands of meters).

Refraction Microtremor (ReMi) techniques (Louie, 2001) traditionally rely on linear arrays and can consequently have some problems related to the directionality of the events. Being a technique based on the analysis of seismic events from

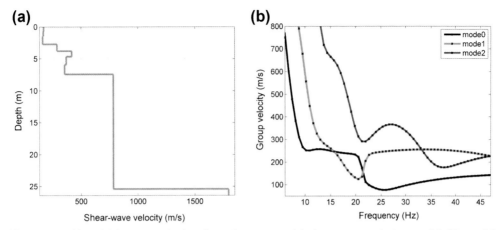

Figure 3.14 Nontrivial group-velocity dispersion curves: (a) shear-wave velocity model, (b) modal dispersion curves for the first three modes. In the 16–21 Hz frequency range, the first higher mode is slower than the fundamental one, while in the 33–47 Hz interval, the second higher mode is slower than the first overtone.

unknown sources (which can be also perpendicular to the array direction), the iden tification of the effective dispersion curve can be a bit problematic and fuzzy (in any case, it must be considered the "lower bound" that describe the lowest velocity—see Figure 3.15).

Figure 3.16 reports a lucky example for which the dispersion curve retrieved from an active acquisition (MASW) results in continuity with the ReMi velocity spectrum.

Anyway, as for the active acquisitions (considering smaller offsets), the apparent dispersion curves are always a problem because there is no reason to consider the effective dispersion curve obtained from passive acquisitions (ReMi, Extended Spatial

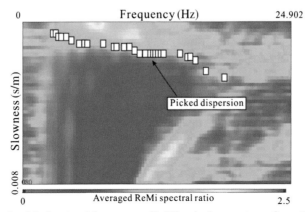

Figure 3.15 Example of Refraction Microtremor (ReMi) velocity spectrum from Cha et al. (2006).

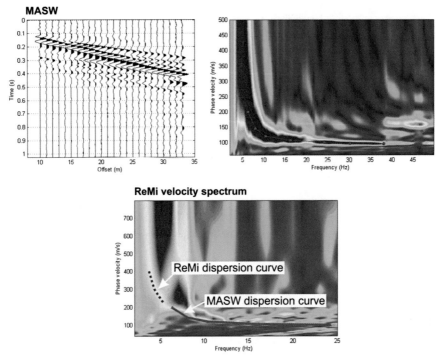

Figure 3.16 Multichannel analysis of surface waves (MASW) and Refraction Microtremor (ReMi) velocity spectra for the same site: the dispersion curve picked while considering the MASW acquisition down to a frequency of about 6 Hz, continues to lower frequencies on the ReMi velocity spectrum.

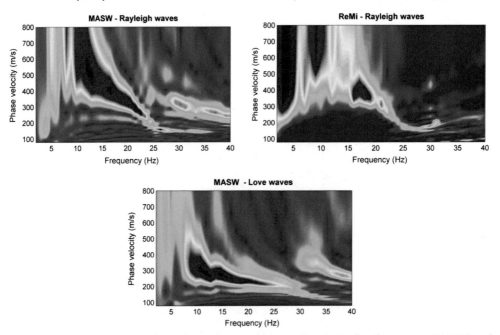

Figure 3.17 Phase-velocity spectra obtained via multichannel analysis of surface waves (MASW) and Refraction Microtremor (ReMi) analyses: in both cases, Rayleigh waves do not show any clear evidence of the fundamental mode (which can be actually revealed through an appropriate data processing). For details see Dal Moro (2011).

Autocorrelation Method (ESAC), or anything else) as belonging to a single mode. Figure 3.17 reports a case where the velocity spectra from both active and passive acquisitions actually show clear evidence of the higher modes only (fundamental mode is present only at very high frequencies (see Dal Moro, 2011)). Figure 3.18 reports an example of ESAC velocity spectrum: the apparent dispersion curve (the yellow line) appears continuous but, in spite of this, it actually pertains to two different modes (see case study 12).

A critical issue in passive measurements can be represented by the dimensions of the geophone array with respect to the *bedrock* (meaning by that, a horizon where a remarkable increase in V_S occurs). Array dimensions must be consistent with it and, as a simple rule of thumb, if you expect a *bedrock* at a depth of about z meters, the dimension of your array (the distance between the two mostly distant geophones) should not exceed more or less three times of z.

Differently than the ReMi approach, the ESAC methodology (Ohori et al., 2002) provides velocity spectra (i.e. apparent dispersion curves) characterized by two points:

1. First of all, thanks to the different mathematics intrinsically capable of handling the directivity of the signals (and that also allows to use bidimensional geophone arrays), the *apparent* (or *effective*) dispersion curve obtained via ESAC is not subject to the problematic "lower bound" issue that instead characterizes any passive method based on a linear array (see and compare Figures 3.15, 3.16 and 3.18 and Paragraph 2.3);
2. Thanks to this, the very low frequencies are better defined.

3.4 FEW FINAL REMARKS

A relevant difference between a multichannel acquisition aimed at defining the phase velocities (MASW, ReMi, ESAC, and similar techniques) and a single- or multichannel acquisition aimed at analyzing group velocities (MFA and related techniques) must be underlined for its practical consequences.

Group-velocity spectra defined through frequency−time analysis (see Section 3.2.2):

1. Individuates the *average* group velocities between the source and the receiver(s) and, consequently, the solution will tell us about the mean V_S model in between.
2. This sort of analyses needs that the *zero time* defined by the seismograph is highly precise.

On the other side, multichannel analysis describing phase velocities according to phase shift or τ-p transforms (e.g., Park et al., 1998; Dal Moro et al., 2003) are based on trace correlations over the considered offsets and not on the absolute values of the surface-wave arrival times.

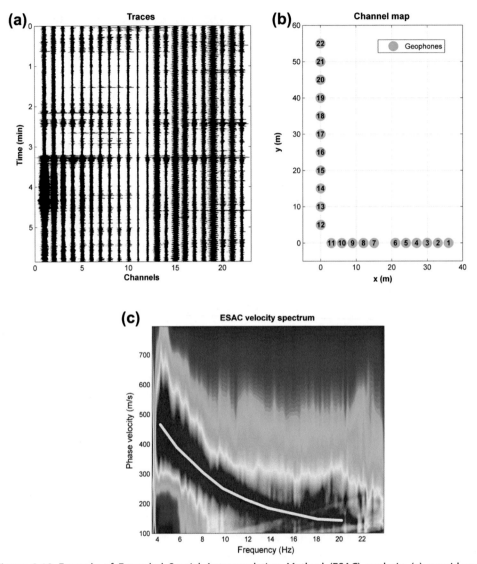

Figure 3.18 Example of Extended Spatial Autocorrelation Method (ESAC) analysis: (a) considered seismic traces (shown only the first 6 min), (b) geophone array (22 geophones), (c) obtained phase-velocity spectrum with, the yellow line, the *apparent dispersion curve*. Please notice that, as for any other (active or passive) method, there is no reason to think that such a dispersion curve is actually related to a single mode (see case study 12).

This means that an imprecise (or completely unknown) zero time is completely irrelevant and this is incidentally the reason why correlation techniques (capable of determining phase velocities) are used also in their passive implementations (ESAC, ReMi, etc.): the arrival time is not relevant but only the differences between the traces.

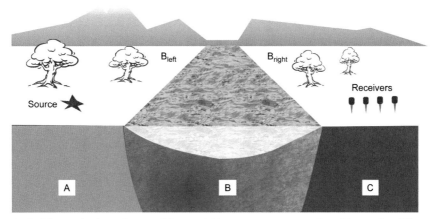

Figure 3.19 Schematic view of a possible acquisition geometry to put in evidence the different practical consequences of frequency–time analyses (such as MFA (Multiple Filter Analysis) or FTAN (Frequency Time Analysis)) and multichannel trace-correlation techniques (such as multichannel analysis of surface waves, Refraction Microtremor, Extended Spatial Autocorrelation Method): the source is on one side of the river while the receiver(s) on the opposite bank. See text.

A practical consequence can be deduced, considering the sketch in Figure 3.19. Let us imagine to set up a seismic source on one bank (B_{left}) of a river and the receivers on the other bank (B_{right}). If we then adopt an MFA (or similar) technique for computing surface-wave group velocities and analyze it, we will eventually end up with a model which is the average model between the source and the receiver(s) including the materials beneath the riverbed.

On the other side, if we use the multichannel data to determine the phase velocities, the model that we will eventually infer from these phase velocities will refer only to the area beneath the deployed receivers (area C).

In the previous sections, we have seen that the radial component of Rayleigh waves is not less important than the vertical one. On the opposite, sometimes it can be even preferable to understand the velocity spectra in terms of modal dispersion curves. A further case for which the radial component can result more important than the vertical one is reported, for instance, by Rodríguez-Castellanos et al. (2006) who, while considering the case of a cracked medium, showed that such a component is capable of giving more emphasis to possible cracks thus resulting quite relevant, for instance, to identify damages on an asphalt cover (Figure 3.20).

So, what is the best component to consider for performing surface-wave analysis? Is the common ZVF component really the best way to consider surface-wave propagation? Or, as we have seen for some cases, the RVF is better? Would that be sufficient?

No cheap and universal answer is viable.

First of all, a single component is rarely sufficient to perform properly constrained analyses hence, multicomponent analyses are always recommended (see Chapters 5

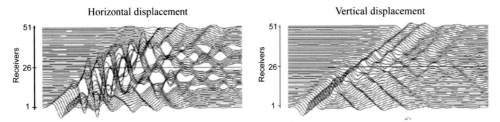

Figure 3.20 Comparison of the synthetic seismograms referred to the radial (left side) and vertical (right side) components of Rayleigh waves in case of surface cracks on a rocky material: radial component results definitely more sensitive to such discontinuities, thus potentially more interesting when the goal is to identify or monitor this sort of features. *From Rodríguez-Castellanos et al. (2006).*

and 6). The "best" components are actually site-dependent and, in order to properly plan an acquisition campaign to explore a large area, a preliminary small survey (just few shots) should be performed in situ. The analysis of these exploratory datasets will provide the necessary insight to understand the complexities of the area and, consequently, identify the most appropriate components to consider to eventually explore the whole area.

In general terms, how should be clear from the synthetic cases illustrated in the Section 3.2.1 and from a number of case studies reported in the Appendix (for instance, case study 4 and 11), thanks to their simple behavior (i.e., dispersion), Love waves are usually a very reliable component that should be considered as a sort of security belt for any kind of survey.

This also means that the very trendy acronym MASW actually means quite little. We can acquire only the vertical component of Rayleigh waves (unfortunately, nowadays this is the most popular way to perform "MASW"), the radial component of Rayleigh waves and/or Love waves, and so forth.

More: once the chosen components are acquired, there are several ways to analyze the data and the Chapters 5 and 6 (together with all the presented case studies) will try to show different possible approaches.

As always, the right approach is not universal and the inversion procedure actually depends on both the goals of the survey and the complexity of the data.

A final important note to be underlined about the higher modes is that they are not a problem in themselves. They can become a problem when misinterpreted, otherwise, they are a very important information to fully (and properly) exploit while inverting the data (see Paragraph 1.6 and Chapter 5). Actually, avoiding analyses based on the *modal* dispersion curves and opting for a more comprehensive approach that considers the *effective* dispersion curve(s) (possibly of multi-component data), it is possible to overcome all those problems described in the previous paragraphs and related to the interpretation of the velocity spectra in terms of modal dispersion curves.

CHAPTER 4

Horizontal-to-Vertical Spectral Ratio

There are more things in heaven and earth, Horatio, Than are dreamt of in your philosophy.
William Shakespeare, Hamlet

4.1 INTRODUCTION

Natural phenomena (such as micro-earthquakes, air-pressure variations connected to meteorological/climatic factors, oceanic waves, etc.) and human activities (industrial activities, traffic, etc.) necessarily produce seismic waves which create a sort of seismic background field.

The term *microtremor* is often correctly used to designate these signals while, as also properly pointed out, for instance, by Cho et al. (2013), we should be more careful while using the expression "ambient noise".

In fact as also described in Chapter 2, the term *noise* is used to indicate any sort of signal (event) that we would not like to be present in the data. It is therefore fully illogical (an absolute contradiction) to define as *noise* something that we are actually seeking, analyzing, and using.

The knowledge of the microtremor background (and its exploitation) dates back at least to the 1950s (e.g., Gutenberg, 1958): oceanic/sea waves, winds, local and regional meteorological perturbations, tectonic and volcanic microseismicity, and industrial (or, in general terms, artificial) activities produce body and surface seismic waves that travel along the Earth surface thus determining a microtremor field particularly important in the low-frequency range (e.g., Cessaro, 1994; Koper et al., 2010) (Table 4.1). Of course it should be considered that the attempt to classify and limit the influence of a specific component/source into a strictly defined frequency range must be regarded with some flexibility since it would be clearly naive (or psychotic) to think that the complexity and variability of the natural phenomena can be fixed within a rigid grid we usually adopt to describe and assess the universe, and the word to the wise of Hamlet to Horatio (the opening quote of this chapter) should be always remembered.

The intuition (clearly also supported by theoretical considerations) that the spectral ratio between the horizontal and vertical components of the background microtremor field (HVSR—*horizontal-to-vertical spectral ratio*) could be used for assessing the site amplification in case of earthquake was notoriously put forward by the Japanese researcher Nakamura since the 1980s (1989; 1996; 2000).

Surface Wave Analysis for Near Surface Applications
ISBN 978-0-12-800770-9, http://dx.doi.org/10.1016/B978-0-12-800770-9.00004-2

Table 4.1 Microtremors: Natural and Artificial Components in Different Frequency Range According to Classical (Early) Studies

	Gutenberg (1958)	Asten (1978) and Asten and Henstridge (1984)
Sea waves	0.05–0.1 Hz	0.5–1.2 Hz
Large-scale meteorological perturbations	0.1–0.25 Hz	0.16–0.5 Hz
Oceanic cyclones	0.3–1 Hz	0.5–3 Hz
Local meteorological perturbations	1.4–5 Hz	–
Volcanic tremors	2–10 Hz	–
Urban (artificial) components	1–100 Hz	1.4–30 Hz

Of course, although we will briefly underline few key aspects related to the evaluation of the meaningfulness of the HVSR curves/peaks, here we are not going to discuss the aspects related to the role of HVSR in seismic hazard studies, and only focus on some critical aspects related to its use (and sometimes abuse) in shear-wave velocity determination.

4.2 DATA ACQUISITION AND HVSR COMPUTATION

4.2.1 Data Acquisition

The *horizontal-to-vertical spectral ratio* is computed by considering the data acquired using a properly calibrated 3-component (3C) geophone (please notice that some authors and some companies use the expression 3D instead of 3C). Even if we are not going to discuss excessively technical hardware aspects, it is anyway important to underline that, in this case, by *calibration* we mean the fact that the three geophones (two horizontal and perpendicular to each other and a vertical one) must have exactly the same response curve all over the frequency range we are interested in.

As taught in any basic geophysical course, by *geophone response curve* it is meant the output of the geophone (typically expressed in V/m/s or V/cm/s or, in the several Anglo-Saxon countries, as V/in/s) as a function of the frequency.

That simply means that if we have two identical vibrations purely along the x-axis and z-axis, the electrical output generated by the x and z geophones must be the same. This way we will be able to properly analyze the amplitude ratios that, in case the two geophones would not have the same response curve, would be otherwise impossible to retrieve.

On the other side, by *equalization* it is usually meant that operation capable of modifying the amplitude of the geophone output, generally with the aim of compensating the minor sensibility of the high-frequency geophones in the low-frequency range (this is possible only if the *response curve* of the geophone is known).

That means that if, for academic reasons, we are interested in the evaluation of the amplitude spectra of the single components we must deal with data acquired with very low-frequency sensors (commonly adopted in seismological studies) or, in a cheaper perspective, with properly equalized data.

On the other hand, if (for practical purposes) we are merely interested in the HVSR curve, it is only necessary to deal with calibrated sensors.

A number of documents describing the procedures to adopt for acquiring data to use for the HVSR computation are available (see SESAME project—AA.VV., 2005) but it is probably useful to summarize here the critical aspects.

Being a passive acquisition, the only parameters to fix are the sampling rate and the acquisition time (or length). As always, the values of these two parameters cannot be fixed on a universal perspective since they depend on the goal of the survey and local stratigraphy.

In general terms it is anyway important to recall that the sampling rate depends on the maximum frequency we are actually interested in. The Nyquist frequency (which is simply half of the sampling frequency) will be then fixed according to this. If, for instance, we want to analyze frequencies up to 30 Hz, it will be sufficient to adopt a 60 Hz sampling frequency (i.e., 0.0167 s sampling interval).

It is then clear that the sampling frequency to adopt is a function of the goal and site stratigraphy, since the answer to the question "*how deep do you want to go*" depends on both these two aspects in a way that will be clarified later on in this chapter. In general terms it can be anyway said that—for geological purposes—a 128 Hz sampling frequency (often the sampling frequency is a power of 2) is absolutely more than sufficient.

The second parameter is the recording length: how many minutes should we acquire in order to obtain a sufficient amount of data for the statistical analyses performed in order to determine a robust HVSR curve?

The common (cheap) answer to this question is 20 min, but this figure can result inadequate if a number of aspects related to the nature of the microtremor field and the statistical criteria to meet are not properly considered.

One of the SESAME criteria defining the overall meaningfulness of determined HVSR curve imposes that the number of significant cycles (indicated as nc) must be higher than 200. Since nc depends on several parameters (namely the peak frequency and the number and length of the analysis window) it is not possible to provide a simple and universal answer to this point. A rule-of-thumb table was anyway proposed (AA.VV., 2005) and briefly reported in Table 4.2. The point can be summarized as follows: if you suspect a very low-frequency HVSR peak (around 0.2 Hz) you must acquire at least half an hour of data and, during the analysis, apply a window analysis (l_w) of minimum 50 s while, on the opposite extreme, if you are only interested in a possible peak at about 10 Hz, the window length (to fix during the analysis) can be of just 5 s and the recording length of just 2 min.

Table 4.2 Recommended Acquisition and Processing Values for Horizontal-to-Vertical Spectral Ratio Analyses

f_0 (Hz)	Minimum Value for l_W (s)	Recommended Minimum Record Duration (min)
0.2	50	30
0.5	20	20
1	10	10
2	5	5
5	5	3
10	5	2

f_0, the peak frequency (which depends on the site) and l_W, the length of the window (to fix during the processing).
After AA.VV. (2005).

Since typically we do not know in advance where the HVSR peaks will occur, Table 4.2 must be read from a different perspective: down to which frequency do we want to go? If we are interested in frequencies down to 1 Hz it is recommended to acquire at least 10 min of data; instead if we want to acquire data to analyze down to 0.2 Hz it will be necessary to acquire at least half an hour of data.

Actually, as we will see in the next paragraph, because of possible meteorological/climatic influences, things can be more complex and multiple acquisitions (on different days) can be sometimes useful or necessary.

4.2.2 HVSR Computation

Figure 4.1 reports an example of dataset useful for HVSR analyses: a three-component 20 min recording. Three rectangles put in evidence three transients (large-amplitude events that can be related, for instance, to a passing-by lorry or a gust of wind).

In the classical approach (or theory), it is recommended to remove these transient events because their H/V spectral ratio can be different/alien with respect to the natural one, representative of the site. Nevertheless, it can be argued that if such signals relate to distant events and are not too large often their H/V spectral ratio will not significantly alter the final HVSR (which—as shall be shortly seen—is an average value). Furthermore, some researchers (e.g., Mucciarelli et al., 2003; Parolai and Galiana-Merino, 2006; Parolai et al., 2009) somehow propose exactly the opposite approach: these large-amplitude transients can be more representative of the site behavior in case of earthquakes. While this perspective can be valid when we consider the HVSR curve in terms of seismic hazard studies, it can be anyway underlined that the necessary preprocessing (removing or not the transient events) clearly depends on the use that will be done of the retrieved HVSR curve. If the goal is to support our analyses aimed at retrieving the V_S profile, it must be underlined that the HVSR modeling is a complex issue that

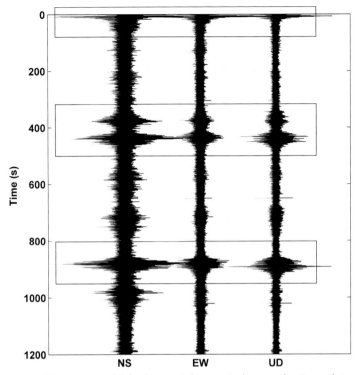

Figure 4.1 Example of datasets useful for horizontal-to-vertical spectral ratio analyses: three rectangles highlight three large-amplitude events (typically named *transient*) that we can decide to remove before the computation of the H/V spectral ratio. UD stands for *up–down* (i.e., the vertical component), EW for East-West and NS for North-South.

should be performed while properly considering the contribution of both Rayleigh and Love waves (e.g., Arai and Tokimatsu, 2004; Lunedei and Albarello, 2009). In that case the removal of events not pertaining to the background steady field of microtremors can be probably preferable.

Once transient events are removed, three fundamental parameters must be fixed: window length (l_W), spectral smoothing, and tapering.

As previously mentioned, the length of the window (see Figure 4.2) should be a function of the frequency peak(s) we want to highlight (see Table 4.2) since we need to fix such a length in order to have a minimum amount of cycles within it. The window (whose length can be, for instance, of 20 s) is then moved along the dataset and the H/V spectral ratio is computed for all the considered windows (i.e., segments). Finally, all the computed HVSR curves (for all the considered windows) will be averaged to obtain a mean curve and the consequent standard deviations. HVSR is thus a statistical curve and its validity will depend on the amount (and quality) of considered data and, clearly, on the adopted parameters. With respect to the way we decide to move the

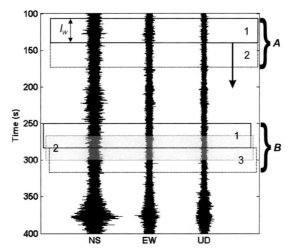

Figure 4.2 Horizontal-to-vertical spectral ratio (HVSR) computation: a window (whose length l_W is fixed by the user—for typical values see Table 4.2) moves along the dataset. HVSR curve is defined within each window and eventually averaged to obtain a mean HVSR curve, together with its standard deviation. See text for further details. UD, up–down.

analysis window along the traces in order to compute all the HVSR curves eventually averaged to obtain the final mean HVSR curve, a final point can be considered. In Figure 4.2 two possible strategies are illustrated. For the first approach (labeled as *A*) the beginning of the second window is set just at the end of the first one, while for the second approach (labeled as *B*) there is a 50% overlapping in the data windowing. The second approach is clearly going to increase the number of windows (thus somehow improving the statistical properties of the retrieved HVSR mean curve) also exploiting that part of the data which is removed by the tapering (see later on).

The spectral smoothing (applied to get smoother amplitude spectra and consequently HVSR curves) is a further parameter that can result relevant especially when we are interested in putting in evidence low-frequency peaks. An example of HVSR processing with some comments aimed at putting in evidence this aspect is reported in the Appendix (case study 13).

Few words about *tapering* are finally necessary. This is a well-known operation performed in a myriad of signal-processing operations in order to avoid problems related to the signal truncation since, if not properly smoothed out, the edges of a truncated signal can create artifacts during the computation of the Fourier spectra. The concept is schematically represented in Figure 4.3 where, from top to bottom, are reported the data of the original signal (whose length—see Figure 4.2—is l_W), the tapering function (commonly expressed as a percentage of the total length of the segment) used to smooth the edges, and, finally, the signal obtained by the multiplication of the original signal by the tapering function.

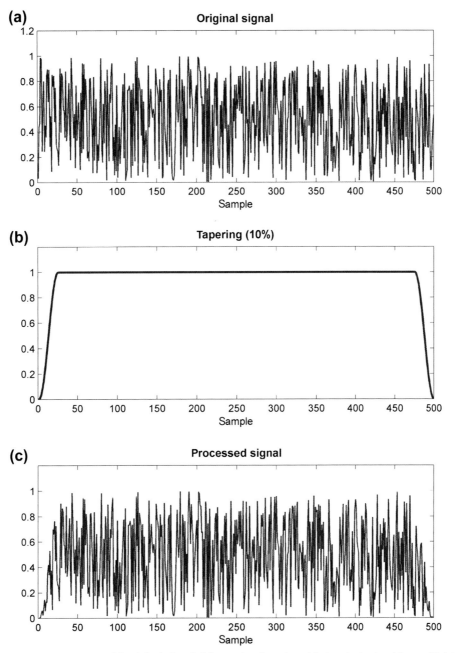

Figure 4.3 Data tapering: (a) original signal; (b) tapering function; (c) signal obtained by multiplying the original signal by the tapering function.

4.3 SOME PROBLEMS

In the previous chapter, together with the opportunities offered by the surface-wave analysis, a series of problematic aspects related to their propagation (and dispersion) were also highlighted eventually underlying that only the analysis of the effective dispersion curve(s) of multi-component data can actually solve all the possible ambiguities otherwise affecting the data (a number of concrete case studies are presented in the Appendix).

As highlighted by several researchers, being the result of passive measurements, HVSR curves represent the combination of several necessarily uncontrolled factors: the relative amount of body and surface waves, the relative ratio between Rayleigh and Love waves, the variability of the sources responsible for the observed HVSR, etc.

> *Contrary to a dispersion curve, if the amplitude is slightly biased you may get very different velocity profiles. Inverting ellipticity curves can be only viewed as an add-on to array measurements or when V_S is already known by other methods. Keep in mind that H/V amplitudes depend strongly on the measurement conditions. The amplitudes can vary along with time at the same site (hence for the same structure). Inverting this curve is simply NOT reliable unless you perfectly know the wavefield contents at the time of the experiment (rather impossible nowadays).*
>
> **Marc Wathelet, from the forum of the SESAME project.**

In the next pages we shall briefly describe the main issues (summarized by Wathelet) that must be kept in mind while analyzing the H/V spectral ratio.

4.3.1 Microtremors and Meteorological/Climatic Variations

One of the first possible problems to mention is probably the one connected to the effects of meteorological/climatic variations on the observed HVSR curves.

In the previous paragraphs we briefly recalled the fact that the microtremor field is generated by a number of natural (and artificial) events and phenomena (see Table 4.1) that somehow create a kind of vibrating surface.

At least to some degree, a variation in the dominating sources (in their nature and amplitude) will necessarily reflect in a different HVSR curve. We should in fact imagine the Earth as a *system* and the observed HVSR curve as the effect (i.e., the output) of the *Earth system* on the input signal provided by the sources briefly recalled in the introductory paragraph.

How large can be the difference in the HVSR curves determined (clearly on the same point and by using the same equipment) in two different days/seasons?

Quite clearly there is no universal answer to this dilemma and the only possible thing is to show few examples that should suggest some caution while evaluating a HVSR curve.

Figure 4.4 shows the amplitude spectra of the three components recorded before, during, and after the passage of a hurricane in the United Arab Emirates. Apart from

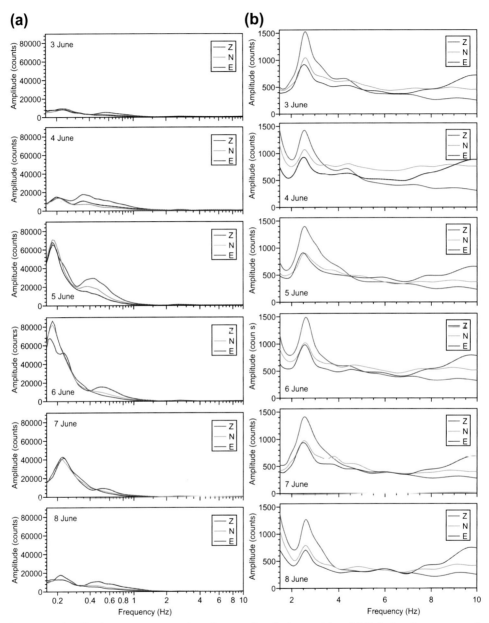

Figure 4.4 Amplitude spectra comparison for two sites (columns (a) and (b)) during the passage of a hurricane. *From Ali et al. (2009).*

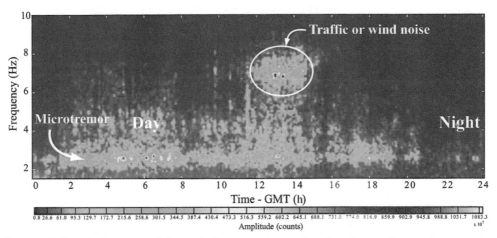

Figure 4.5 Amplitude spectra of the vertical component as a function of time. Changes between the diurnal and nocturnal hours appear straightforward. *From Ali et al. (2009).*

the apparent amplitude variations, it is important to point out the fact that the relationships (thus the ratios) among the different components change and, consequently and necessarily the H/V spectral ratio as well.

The same authors also show the difference between the night and day regimes (Figure 4.5) very likely related to the influence of artificial activities (absent during the nocturnal hours) also maybe jointly with the influence of the wind.

A couple of dramatic examples of variability in the HVSR curves obtained (on the same point) under different meteorological/climatic conditions are reported in Figures 4.6 and 4.7.

In both cases the differences are quite remarkable and should represent a warning for all those practitioners and researchers who intend to use HVSR in their work.

It can be incidentally underlined that, in this context, amplitude spectra and HVSR curves are typically presented while adopting logarithmic scales (remember that $10^0 = 1$).

Further examples and studies presenting the effects of different seasonal or meteorological effects are presented for instance in Tanimoto (1999) and Tanimoto et al. (2006).

4.3.2 HVSR, V_S Profile, and Nonuniqueness of the Solution

Although, as previously recalled, the meaning of the H/V spectral ratio in the framework of seismic hazard studies became clear only in the 1980s; its use in terms of support in the determination of the V_S vertical profile is even older. For instance, Mark and Sutton (1975) used the H/V spectral ratio to estimate the shear-wave velocity profile of some lunar landing sites (see also case study 14).

The problem we intend to briefly recall in this section was already faced in Chapter 3 with respect to the surface-wave analysis: the nonuniqueness of the solution or, in other

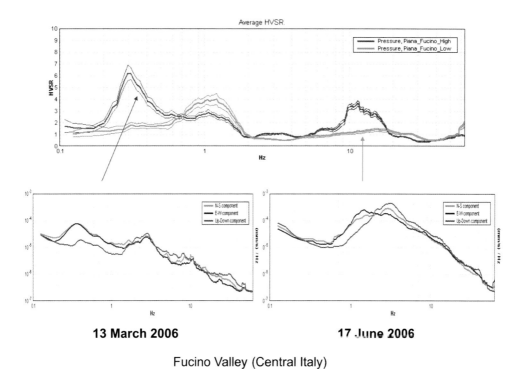

13 March 2006   **17 June 2006**

Fucino Valley (Central Italy)

Figure 4.6 H/V spectral ratios observed on the same site under different meteorological/seasonal conditions. *From Albarello (2006).*

but equivalent terms, the equivalence of the subsurface models capable of describing an observed HVSR curve.

Figure 4.8 reports an example of such equivalence severely affecting the H/V spectral ratio with respect to the attempt of using it to infer the V_S profile: three V_S models are reported in the upper panel, while in the lower one their H/V spectral ratios (computed according to Lunedei and Albarello, 2009) are reported. As can be seen, three completely different V_S models produce three nearly identical HVSR curves (two curves are actually absolutely identical and fully overlap, while the third one is just slightly different). It must be also considered that, in the real world, HVSR curves also have some uncertainties and ambiguities connected to the way they are determined (see previous sections) and some ambiguities are also present in the way HVSR is modeled (see next section).

Of course, since HVSR and surface wave dispersion are two different so-to-say phenomena (even if, to some degree, they are different aspects of the same thing—i.e., the surface-wave propagation), there is a difference in their nonuniqueness.

The main point characterizing these two forms of nonuniqueness can be pointed out simply comparing the plots presented in Figures 4.8 and 1.17.

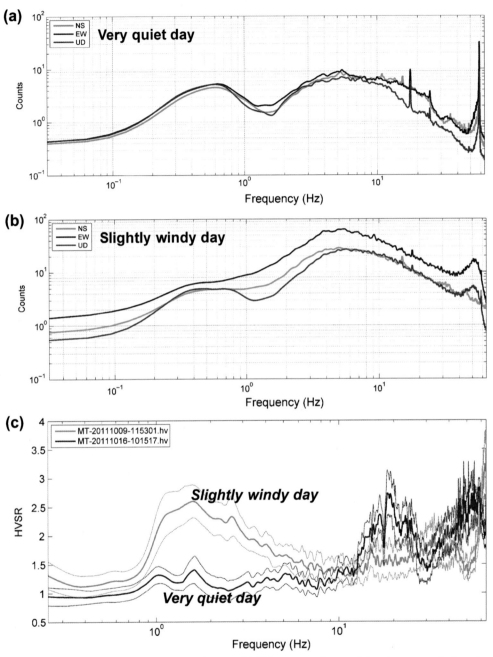

Figure 4.7 Horizontal-to-Vertical Spectral Ratio (HVSR) curves observed (at the same site) under different meteorological conditions: (a) amplitude spectra of the data recorded during a very quiet day; (b) amplitude spectra of the data recorded (few days before) during a slightly windy day; (c) comparison of the two HVSR curves.

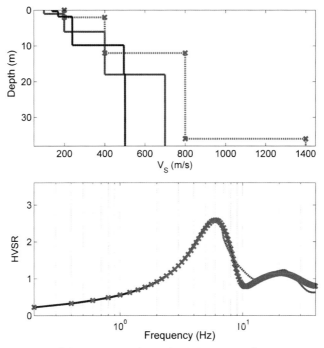

Figure 4.8 Nonuniqueness of the H/V spectral ratio (HVSR) in terms of reconstruction of the V_S profile.

HVSR is sensitive to shear-wave impedance contrasts but can be used to identify the shear-wave velocities only if the depths/thicknesses of the local stratigraphy are known (or, clearly, vice versa). On the other hand, surface-wave dispersion is somehow short-sighted and, below a certain depth (which depends on the combination of adopted sensors, length of the array and considered seismic source), it is not capable of identifying the correct model even if, when the velocity spectra are properly processed, down to a certain depth the solution is not affected by any ambiguity (see Paragraph 1.6).

4.3.3 Attenuation, Modes, Topography, Industrial Peaks, and Asphalts

In this section we shall face the main aspects related to the HVSR modeling we might decide to perform (never forgetting about the previously mentioned issues) in order to get information about the local V_S vertical profile. It will be necessarily impossible to present all the details but it is nevertheless important to fix the most relevant points.

The classical (Nakamura) theory would suggest that the driving force that determines the observed HVSR is represented by the body waves, but more recent studies indicate that what really counts are the surface waves (e.g., Fäh et al., 2001; Bonnefoy–Claudet et al., 2006, 2008).

Actually, even assuming this point as clarified and that in the microtremor background field surface waves dominates over body waves (actually under some geological/stratigraphic circumstances the so-called fundamental resonance frequency can be better explained in terms of body waves rather than by means of surface waves—see e.g., Albarello and Lunedei, 2010; Dal Moro, 2010), a number of further issues necessarily arise: How many modes are relevant? How important is the attenuation in the observed H/V Spectral Ratio? What is the relative contribution of Rayleigh and Love waves in determining the experimental field HVSR?

About the relative contribution of Rayleigh and Love waves in the background microtremor field (e.g., Arai and Tokimatsu, 2004; Bonnefoy-Claudet et al., 2008), this is clearly a general issue with, as usual, no universal answer. Here it is just important to underline that in the observed HVSR curve the contribution of the Love waves is decisive. Unfortunately, in the literature it is not uncommon to read that HVSR is the effect of Rayleigh-wave ellipticity, but this sort of statements are actually not only simplistic but profoundly erroneous.

If we accept and assume that the microtremor field is purely composed of surface waves, the horizontal component of the ground motion is actually influenced by both Love and Rayleigh waves, while the vertical one is determined by Rayleigh waves only, which—different from Love waves—have both vertical and horizontal (actually *radial*—see Chapter 1) components. That means that, in general terms, the larger the amount of Love waves in the background microtremor field, the larger the values of the observed HVSR.

The relative amount of Rayleigh and Love waves is clearly a site- and probably time-dependent parameter and it is only for the need of reducing the number of variables in the problem that it is sometimes assumed that the ratio between Rayleigh- and Love-wave contribution is 0.7 (e.g. Arai and Tokimatsu, 2004).

An example will put in evidence the joint influence of the number of modes and attenuation on the H/V spectral ratio. Figures 4.9 reports a V_S model and, in the lower panel, the HVSR curves computed (according to Lunedei and Albarello, 2009) in one case (black continuous line) considering the pure elastic case and only one mode, while in the second case (magenta dashed line) three modes were used, jointly with some attenuation as well.

The difference between the two curves is noticeable and should, together with all the other evidences, represent a warning for any simplistic assumption while modeling the H/V spectral ratio (for a further example of modeling aimed at highlighting the role of the number of modes and attenuation, see Dal Moro, 2011).

In general it is important to point out that the modeling of the HVSR according to surface waves requires to account for several modes (one is rarely sufficient) and attenuation (the unrealistic elastic case risks to be inadequate). An interesting case study providing great evidence of the possible role of attenuation in the observed H/V spectral ratio is reported in the Appendix while considering the joint analysis of some seismic data acquired on the Moon during the epic Apollo missions (case study 14).

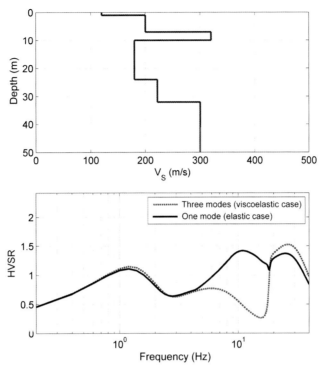

Figure 4.9 Role of the number of modes and attenuation on the H/V spectral ratio while considering the formulation of Lunedei and Albarello (2009). The black continuous line represents the horizontal-to-vertical spectral ratio (HVSR) computed while considering only the fundamental mode in an elastic perspective while the dashed magenta line represents the HVSR obtained considering the same V_S model (reported in the upper panel) but now considering three modes and some attenuation.

In can be pointed out that, similarly to what we have already seen about the excitation of higher modes in surface wave propagation (Chapter 3), strong and shallow shear-impedance variations (due, for instance, to the contact between a superficial soft-sediments cover and a gravel or conglomerate layer) emphasize the role of higher modes also in the HVSR curves (for some details about the role of attenuation on HVSR see e.g. Lunedei and Albarello, 2009).

Finally, Figure 4.10 reports a further example of HVSR computation done considering the body-wave formulation (Herak, 2009) and the surface-wave approach (Lunedei and Albarello, 2009). It is possible to point out how the two different approaches are similar in the low-frequency range but significantly differs at higher frequencies. Non-flat topography can also create H/V peaks that are not connected to stratigraphic components (as a matter of fact, a topographic relief is a structure and as such can vibrate for itself). Although the relationship between the possible directivity (i.e. the variation of the HVSR curves with respect to the azimuth) of the signal and the topography can

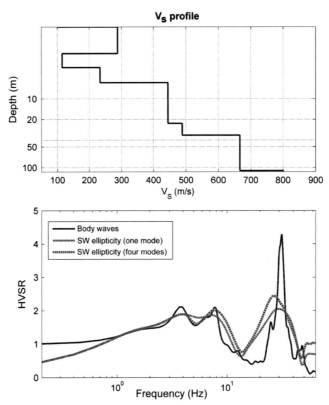

Figure 4.10 Horizontal-to-vertical spectral ratio (HVSR) modeling for the V$_S$ model reported in the upper panel (please notice that—in order to better visualize the shallow part of the model—the y-axis is logarithmic). In the lower panel are the H/V spectral ratios obtained while considering body waves (Herak, 2008) or surface waves (Lunedei and Albarello, 2009) in case of one or four modes.

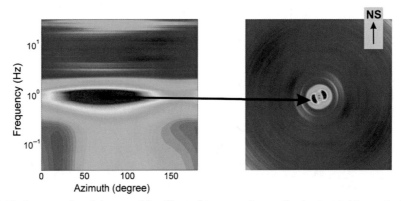

Figure 4.11 An example of the possible effect of topography on the horizontal-to-vertical spectral ratio data: the peak at about 1 Hz shows a strong directivity in the N70E direction, which corresponds to the maximum slope.

be notably complex and consequently the link quite vague, Figure 4.11 represents a typical case for which the directivity is clearly related to the slope since in that case the HVSR peak around 1 Hz shows a dramatic maximum in the N70E direction, which in this cases coincides with the slope. For these reasons, while performing acquisitions over a slope it is often a good practice to fix the instrumental North—South direction not according to the geographical North but along the maximum slope: this way the possible connection between slope and directivity will be even more straightforward (see case study 2).

What is the influence of the human (industrial) activities on the observed HVSR? Often their appearance on the amplitude spectra and on the consequent HVSR curve is quite simple to spot. An example will help illustrate the two main facts that must be considered in order to understand whether a signal (a peak in the H/V spectral ratio) is related or not to an artificial component (not connected to a stratigraphic feature).

Figure 4.12 reports the amplitude spectra and the HVSR for a dataset collected in NE Italy. The signal labeled as B represents a stratigraphic feature (HVSR peak) for at least two reasons:

1. The amplitude spectrum of the vertical component (red curve in the upper panel) decreases with respect to the two horizontal components, thus forming a sort of elliptical shape (ogiva-like) highlighted by the light-pink color (if the amplitude of the vertical component is smaller than the horizontal one, a peak in the H/V ratio will clearly form).
2. It occurs along a wide frequency range (in this case between about 4 and 20 Hz).

An industrial peak (in general, we should speak about "artificial" signal due to any sort of human activity) is differently characterized (see the signal labeled as A in Figure 4.12):

1. It spans along a much smaller frequency range (please remember that the frequency axis is logarithmic—in this case the signal/peak is bounded in the 0.8—1.3 Hz frequency range).
2. It is typically created by an increase of all the three components but very often (and this is one of these typical cases) the increase in the vertical component is smaller than the peak in the horizontal components consequently creating a peak in the H/V spectral ratio.

This HVSR curve actually deserves two further comments:

1. In the 0.6—2.2 Hz frequency range the industrial peak (A) actually overlaps with a stratigraphic one (signal C partially reconstructed/interpreted by the green dashed line).
2. In the 2.2—4.5 Hz frequency range (see the peak labeled as D) the H/V spectral ratio is significantly and firmly smaller than 1 and although it is always quite dangerous to put forward generalizations universally valid, this kind of feature (H/V ratio smaller than 1) can represent a clue suggesting the presence of a low shear-wave velocity layer (see later on what happens to the horizontal component when a shear-wave velocity inversion occurs).

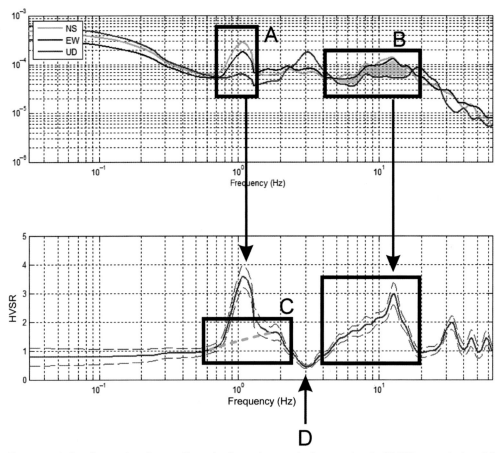

Figure 4.12 Reading and understanding a horizontal-to-vertical spectral ratio (HVSR) curve: industrial (artifical) peak (signal labeled as A) overlaying a stratigraphic one (signal C with the interpolated/interpreted signal related to the purely stratigraphic signal indicated by the dashed green line). The B peak is purely stratigraphic. The signal D (HVSR values significantly and consistently smaller than 1) indicates a likely V_S inversion. For further comments, see text. UD, up–down.

What happens if the 3C geophone lies over a stiff superficial layer (such as an asphalt cover) instead of being posed on a soft cover? The effect is that possible high-frequency peaks will be smoothed out thus somehow altering the "real" HVSR otherwise determined by the underneath stratigraphy.

Once again, an example will clarify this point. Figure 4.13 reports two datasets collected on the same spot (an alluvial plain characterized by soft sediments) just a couple of meters away one from the other. The data in the upper panel (Figure 4.13(a)) represents the acquisition done over the asphalt cover (beneath the asphalt was present a 30 cm gravel-like stratum that, overall, created a stiff superficial layer and consequently a

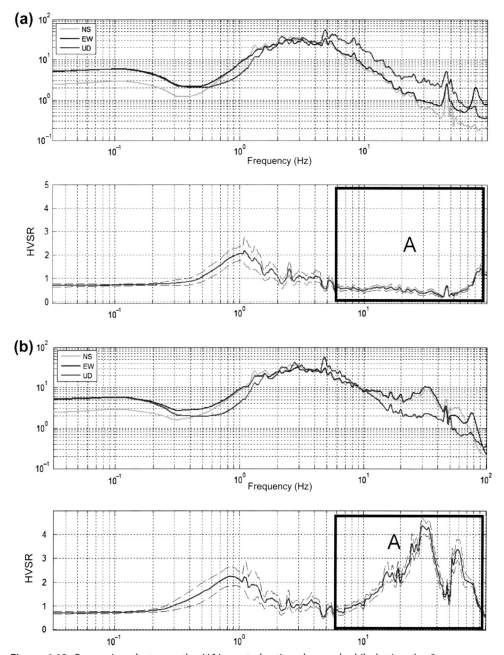

Figure 4.13 Comparison between the H/V spectral ratios observed while letting the 3-component geophone (a) on an asphalt cover and, just a couple of meters away from it, (b) on the natural soil. See text. HVSR, horizontal-to-vertical spectral ratio.

velocity inversion) while the lower panel reports the data collected fixing the 3C geophone directly on the soil. For frequencies higher than about 8 Hz, the stiff cover determines a severe decrease of the horizontal components (compare the amplitude spectra of the two datasets) and, as a result, a strong decrease in the H/V spectral ratio (compare the HVSR curves within the two "A" boxes). When a V_S inversion happens in depth, the consequence is that the H/V spectral ratio tends to decrease and can reach values minor than one (see the "D" signal in Figure 4.12 and the related text).

The consequences in case of measurements performed in the framework of seismic hazard studies should be apparent (possible high-frequency resonances are in fact completely smoothed out) and care must be then paid when it is not possible to work directly on the natural soil.

The effect of Love waves on the *Horizontal-to-Vertical Spectral Ratio*: practical consequences

As previously mentioned, the HVSR determined from field data is the result of the combined effect of both Rayleigh and Love waves, basically according to the following equation:

$$HSVR(f) = \sqrt{\frac{\alpha H_L(f) + H_R(f)}{V_R(f)}}$$

where H_R and V_R are the Rayleigh-wave contributions (in terms of power spectra—see Arai and Tokimatsu, 2004) on the horizontal (H) and vertical (V) axes, while H_L relates to Love waves (the α parameter can be considered as the amount of Love waves in the background microtremor field).

While some authors forget to include the effect of Love waves and treat the H/V spectral ratio as simply representative of the Rayleigh-wave ellipticity, some others properly include their effect in the HVSR modelling. In the latter case anyway, a problem arises about the proper value of the α parameter, which is likely related to the specific overall regime that characterizes the region and the meteorological/climatic conditions. Incidentally, it can be speculated that the differences in the H/V spectral ratios observed in different days/seasons (see Paragraph 4.3.1) could be the result of a different amount of Love waves in the background microtremor field and, for the very shallow part (few meters), the result of different water content with the consequent variations in shear- and compressional-wave velocities (see also paragraph 7.2.2).

An elementary modelling will clarify the point. Figure 4.14 reports a V_S model (upper panel) and the two HVSR curves (lower panel) obtained while considering a small ($\alpha = 0.2$) and a high ($\alpha = 0.9$) amount of Love waves. As can be seen, the obtained HVSR curves are significantly different (since Love waves moves only on the horizontal plane, the effect is an overall increase in the H/V curve).

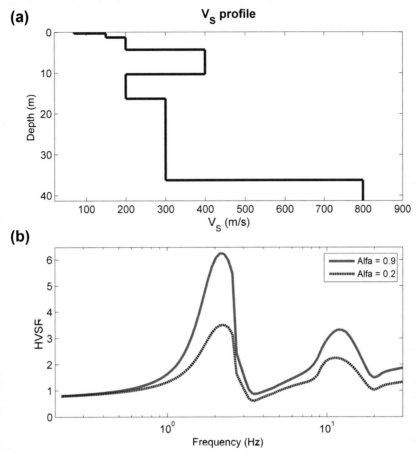

Figure 4.14 Effects of Love waves on the HVSR: (a) considered V_S profile; (b) the HVSR curves obtained while considering a different amount of Love waves (the α factor) in the microtremor field. In both cases Q_S values are fixed according to a simple rule of thumb ($Q_S = V_S/8$).

Two consequences are straightforward:

1. The amount of Love waves (synthetically expressed by the α factor) should be considered as a further variable in the inversion process aimed at determining the V_S profile (experience teaches that its value typically ranges from 0.3 to 0.6).
2. The HVSR curve alone is insufficient to properly and precisely define a V_S profile even when geological/stratigraphical information are available and, consequently, the only viable approach is represented by the joint inversion with further geophysical data (typically the dispersion curves of Rayleigh or Love waves).

CHAPTER 5

Inversion and Joint Inversion

We'll never get rich by hard work. But, we'll never get rich without it.

Robert Fripp

5.1 INTRODUCTION

For several practitioners, the very concept of inversion is often a bit obscure and its principles sound a bit like esoteric formulas hard to catch in their concrete implications. In this chapter, we will somehow try to fill the gap between academic research papers and practical consequences while analyzing field datasets.

Several practitioners in fact, imagine that the *inversion* is that *process* for which data are put into a magic black box and a subsurface model is somehow obtained. Little care is typically paid to the fundamental fact that the quality of the output (the retrieved model) depends on the quality of the input and on the parameters adopted for the inversion process, which must be clearly understood in its basic founding and driving principles.

In Chapters 1 and 3 the basic ideas about the link between a given subsurface model and the dispersion (and attenuation) of the surface waves were illustrated. Fundamentally, the inversion process is the act of reconstruction of the subsurface model from the data acquired on the field.

In our case, the goal is the determination of the V_S vertical profile from the exploitation of the surface-wave dispersion (see Figure 5.1), in some cases jointly with further *objective functions* represented for instance by the refraction or reflection travel times or the H/V spectral ratio.

In general terms, the critical point in this process is that, in order to properly reconstruct the subsurface model, it is necessary to handle a sufficient amount of information: the more information we have, the better the model we can infer. That means that if we do not have a sufficient amount of information our model will be necessarily uncertain or ambiguous or, in some cases, even fully erroneous.

The adopted term (*information*) is actually intentionally vague because the information can be represented both by geophysical data (e.g., seismic data locally acquired) and by supplementary geological information obtained from boreholes or geotechnical investigations, which are typically punctual and, consequently must be considered with some caution when extrapolated to a different point or a wider area.

Since, in order to reconstruct a reliable subsurface model, a minimum amount of information is required, it should be clear that, in general, the more geophysical data we have, the less auxiliary geological information we need.

Surface Wave Analysis for Near Surface Applications
ISBN 978-0-12-800770-9, http://dx.doi.org/10.1016/B978-0-12-800770-9.00005-4

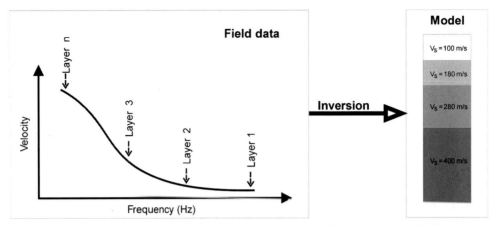

Figure 5.1 Inverting surface-wave data: from the dispersion to the V_S subsurface model. The scheme is about surface-wave dispersion but the principle applies to any physical phenomenon that can be used to infer the properties of the medium.

On the other hand, there are surely cases for which the site (i.e., its stratigraphic sequence) is such that very little effort is necessary to solve it (see for instance the joint group-velocity analyses presented for the case study 2 while considering the data collected by a single 3-component geophone with no ancillary geological information available).

No emphasis was done to the point related to the data quality since it should go without saying that the GIGO acronym that applies to any area for which information is handled, must be always kept in mind: *Garbage In, Garbage Out*.

5.2 MISFIT, INVERSION, AND MODELING: CONCEPTS AND MISCONCEPTS

At its basic level, the concept of *misfit* is quite simple and should be straightforward from Figure 5.2: it is a way to quantify how far is *something* from *something else*.

In this case (the very classical approach to surface-wave inversion) the *misfit* is classically defined as the distance between an observed (which is actually *picked*) dispersion curve and the theoretical one of a tentative model (of course for each considered point there is a *misfit* value and the global one is the normalized value after the summation of them all).

But here we immediately face the first (and probably the most important) problem intrinsic to this classical approach (picking of the dispersion curve to invert by minimizing the *misfit*): in fact what is inverted is not something objective (an observed quantity such might be the air temperature at a given time and space) but rather an *interpretation* (the picking) of the data.

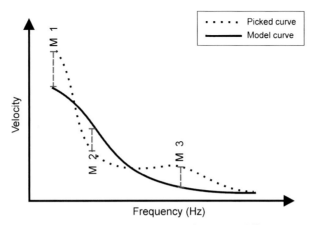

Figure 5.2 While dealing with dispersion curves (in Chapter 6 different approaches are briefly presented), the *misfit* is the summation of the differences between the picked curve and the theoretical one (for a considered tentative model). This value is then normalized by the number of considered frequency—velocity points.

In fact, while the *velocity spectrum* (see Chapter 1) represents an objective quantity (the transformation—performed via uncomplicated mathematical formulas—of the recorded seismic traces into a different and more useful frequency—velocity domain), the picked modal dispersion curves are an interpretation of such a velocity spectrum.

It is actually possible to consider not the modal curves but the "effective" one, but this approach will be briefly illustrated in the next chapter, where some so-to-speak unconventional (and better) methods will be introduced.

Here it is fundamental to underline that (while dealing with modal dispersion curves) the picking represents an interpretation and, for all the reasons illustrated in Chapters 1 and 3, this can be fatally wrong.

The moral is then quite clear: the output of this kind of approach (that we can define as classical) is entirely subject to the correctness of the *picking* (please notice that any form of automatic *picking* of the *modal* dispersion curves is quite clearly highly risky. A further point deserves some clarification.

The *misfit* value obtained while adopting this approach is actually completely meaningless in the sense that it lacks an objective value. In fact, once we pick a dispersion curve we will be surely able to find a model capable of fitting such a curve very well (having, i.e., a very small *misfit* value) even if the picked curve is completely meaningless/wrong. Quite clearly, if we adopt the classical modal dispersion-curve picking, typically the mistake that can be done relates to the misidentification of the correct mode. Evidences of this sort of problems (that can be put in evidence and understood via joint analysis of two or more "objectives") are presented in the case studies 4 and 11.

Once an erroneous *modal* dispersion curve is picked, the way the *misfit* is then minimized within the inversion procedure will not overcome this problem: both traditional methods based on the *Jacobian* matrix (e.g., Xia et al., 1999; Scales et al., 2001) and *global search* methods (e.g., Holland, 1975; Goldberg, 1989; Stoffa and Sen, 1991; Sen and Stoffa, 1992; Smith et al., 1992; Louis et al., 1999; Reeves and Rowe, 2003; Yamanaka, 2005; Dal Moro et al., 2007) will necessarily fail even if a very small misfit is obtained. A straightforward example is provided by the velocity spectrum associated to the ZVF component (the classical way to perform "MASW" analysis by means of acquisitions done while considering only a vertical-impact source and vertical geophones) presented in Figure 3.10. In that case, the fundamental mode is practically completely absent and by picking the first higher mode (which dominates the velocity spectrum) as fundamental mode (mode misinterpretation) we would necessarily end up with a very good match (small misfits between the picked and retrieved modal dispersion curves) but with a fully erroneous shear-wave velocity profile (see also Dal Moro and Ferigo, 2011).

For these reasons, most of the literature focusing on new and fancy inversion scheme is often quite self-referential and has often little practical meaning: once an erroneously picked curve is fed into the fanciest inversion algorithm, the obtained result will be meaningless even if the *misfit* is very small and the difference between the picked and inverted curve is very small.

Things clearly change when we are dealing with a properly picked modal dispersion curve or with an effective dispersion curve (Tokimatsu et al., 1992).

In this case (see next paragraph), compared to the traditional methods based on the Newtonian gradient (e.g., Scales et al., 2001), heuristic methods are capable of better handling the problems related to local minima.

An important point about the *forward modeling* approach is now necessary. So far we have in fact considered the case of an inversion procedure driven by an algorithm that minimizes the *misfit*. Actually a skilled person can adopt a different procedure and decide to personally invert the data. Once the basic notions describing surface-wave dispersion are known, it is actually quite simple to go through a sort of trial-and-error procedure where the user personally modifies the model parameters (in Chapter 1, we have seen that the most important parameters responsible for surface-wave propagation are the shear-wave velocities and the thicknesses of the layers). As a matter of fact the only thing that we must consider while following this procedure is that high frequencies are influenced by the shallowest layers while the deeper we go the lower will be the frequencies that are influenced. In other words, by considering the *steady state approximation* (Chapter 1) and the fact that the higher the V_S the higher the V_R, the practitioner can perform the data inversion without using any automatic procedure capable of finding a mathematical minimum that might have little geological/stratigraphic meaning (see next paragraph). It is in fact possible (and often recommendable) to personally modify V_S or thickness of the layers until a good agreement is found (see the *direct modeling* box). This is particularly useful because this way the user is forced to get familiar with the basic principles that determine surface-wave dispersion and, at the same time, will try to make sense of the inferred shear–wave velocities with respect to the materials (rocks or soils) present in the area. As Stokoe sensibly pronounced during his keynote speech at the Near–Surface EAGE meeting held in Dublin in 2009 (Stokoe, 2009): *We do not invert. We model.*

In a sense, we can thus speak about *automatic inversion* or *direct modeling*, but both these procedures are actually *inversions* because both allow to infer the subsurface model from one or more observed quantities (e.g., velocity spectra, HVSR curves etc.).

Direct Modeling

As illustrated in Figure 5.1, inverting means to determine the subsurface model responsible for the observed data. The *how* is so-to-say secondary and depends on the personal specific needs, attitudes, and knowledge level of the considered phenomenon (in our case surface-wave propagation).

If by "inversion" it is usually intended what, in order to be more precise, could be actually called "automatic inversion" (an automatic procedure that—by minimizing something—will eventually provide a sort of "best model"), a different approach is also possible and often recommendable: the direct modeling of the data, a sort of inversion performed *manually* by the user.

Fundamentally in order to proceed with a direct modeling it is only necessary to consider two facts:

1. Surface-wave velocities (V_R and V_L) are proportional to the shear-wave velocities of the subsurface layers.

2. The rule that connects surface-wave velocities to the subsurface model (fundamentally described in terms of V_S and thicknesses of the layers, with minor contributions from densities, V_P, and quality factors) is represented by the *steady state approximation* (see Chapters 1 and 3) that essentially describes the fact that shallow layers are responsible for the surface-wave velocities at high frequencies while deep strata determine surface-wave velocities in the low-frequency range.

Once these two facts are properly handled, direct modeling becomes an easy task. Figure 5.3(a) reports the phase-velocity spectrum of a field dataset that we might wish to invert by direct modeling. Figure 5.3(b) shows the same data with, overlaying, the theoretical dispersion curve of the fundamental mode for a tentative model chosen (and tested) by the practitioner. It is apparent that the V_S of the model are too high over all the considered frequency range (in this case between 4 and 38 Hz). We will then modify V_S (and/or thickness) values until a good match is reached (Figure 5.3(c)) always simply taking into consideration the $\lambda/3$ rule of thumb described by the *steady state approximation* (so if we observe a mismatch at low frequencies we will modify the deep structure while for the discrepancy at higher frequencies we will modify the shallow layers).

This approach is particularly important for at least two reasons:

• It forces the practitioner to develop a *feeling* of the considered phenomenon (surface-wave dispersion) (if you are not aware of what you are doing, it is quite unlikely that any automatic procedure will help you in getting good analyses).

• It will allow the practitioner to immediately establish a link between the adopted V_S values and the local geological/stratigraphic features.

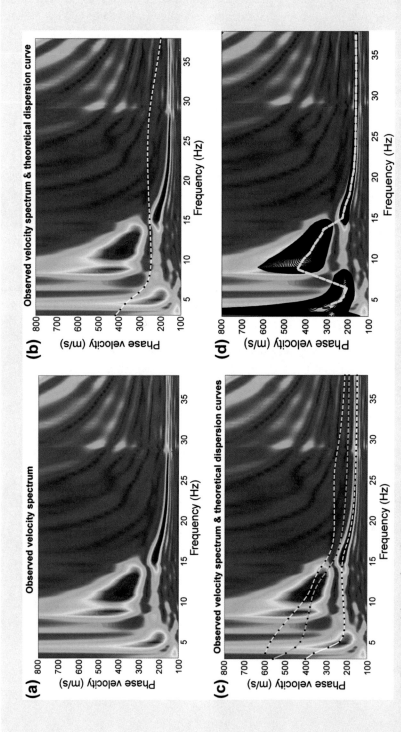

Figure 5.3 Direct modeling of surface-wave dispersion: (a) observed phase-velocity spectrum; (b) computing the theoretical dispersion curve (fundamental mode) of a tentative model (very poor match between the observed spectrum and the theoretical *modal* curve); (c) theoretical *modal* dispersion curves (first three modes) for a model that closely matches the observed dispersion; (d) black contour lines report the phase-velocity spectrum computed from the synthetic seismograms computed from the same model of the previous plot (see *Full Velocity Spectrum* inversion in the next chapter) while the yellow dashed line reports its *effective* dispersion curve.

It is understood that the same procedure can be adopted also while using more datasets (for instance Rayleigh and Love waves and/or horizontal-to-vertical spectral ratio (HVSR) curves) and that different approaches that the one represented by the modal dispersion curves can be adopted. Figure 5.3(d) refers, for instance, to the *Full Velocity Spectrum* analysis described in some detail in the next chapter together with the *effective* dispersion curve and that can be used both in their *automatic* (automatic inversion) and *manual* (direct modeling) forms.

5.3 LOCAL MINIMA AND NONUNIQUENESS OF THE SOLUTION

The main problem with classical Newtonian-based inversion schemes is that they severely suffer from the problem created by the local minima that, so-to-say, attract the model/solution. This is a problem also because these methods usually require a starting model (often fixed according to the *steady state approximation*) and, as a consequence, the final solution will inevitably depend on the adopted starting model.

A graphical representation showing the local-minima problem is reported in Figure 5.4, where the red circles represent different possible starting models that, because of the "Newtonian gravity" simulated by the classical inversion process based on the *Jacobian* matrix, are attracted by the closer local minimum (which represents a "local solution") and stuck there without having the possibility of exploring further minima (i.e., solutions) to identify the *global* one (i.e., the one associated with the *correct* model).

To overcome this intrinsic problem, *global search* methods capable of exploring a user-defined search space without ending up in a local minimum have been proposed, often inspired by the way some biological processes occur.

Some simulations will help visualizing and getting familiar with these aspects (see also Dal Moro and Ferigo, 2011). The Rayleigh-wave fundamental-mode dispersion curve for a *reference model* defined by $V_S = \{150, 280, 150, 400, 600 \text{ m/s}\}$ and thickness THK $= \{2, 4, 5, 7 \text{ m}\}$ was computed in the 5−50 Hz frequency range. Several thousands of different models were considered and the *misfit* between their dispersion curves and the one of the reference model computed.

Figure 5.5 reports the data pertinent to three cases. In the first one, V_{S2} and V_{S3} (the shear-wave velocities for the second and third layer) were set free to vary in the 100−400 m/s range while all the other model parameters were fixed to their real values (i.e., to the ones of the *reference model*). The computed misfits (Figure 5.5(a)) describe a gentle trend with a clear single minimum. Due to the unrealistic nature of such a simulation (all the parameters but two were set to their actual values and no noise was added) the resulting misfit trend would be clearly suitable to be solved even by standard gradient-based inversion algorithms.

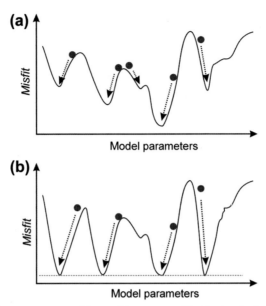

Figure 5.4 Conceptual representation of the problems created by the local minima during the inversion process. While using inversion schemes based on the Jacobian matrix and the gradient approach (e.g., Scales et al., 2011) the final model will depend on the starting model (see upper panel (a)) which—for this kind of methodologies—is mandatory: a model (red circles) is attracted toward the nearest local minimum. The lower panel (b) somehow represents an extreme case for which the four local minima have the same misfit value (blue dotted line) and thus relate to four *equivalent* models.

If we then set free two further variables (the thickness of the second and third layers—THK2 and THK3—are now free to vary in the 2–6 and 3–7 m range, respectively) the system becomes awfully complex because the number of minima and maxima is such that the identification of the *reference model* becomes practically impossible (Figure 5.5(b)). If now we consider the same simulation while jointly considering both fundamental and first higher mode (Figure 5.5(c)), it can be noted that the misfit landscape now becomes more so-to-say characterized (the differences between high and low misfit values increase) thus showing that (when dealing with the modal approach) considering more modes helps in improving the inversion procedure.

The large number of local minima and the complex landscape shown in Figure 5.5(b) and (c) severely challenges any inversion scheme, no matter the principia on which it is based. In other words, the problem (represented also in Figure 5.4) is characterized by such a jagged minima/maxima misfit landscape that the real (reference) model results impossible to differentiate.

The local-minima problem is particularly severe and, also considering the additional uncertainty due to possible "noise" present in the data (e.g., data recorded while considering erroneous or inadequate acquisition parameters, lateral variations, poorly defined

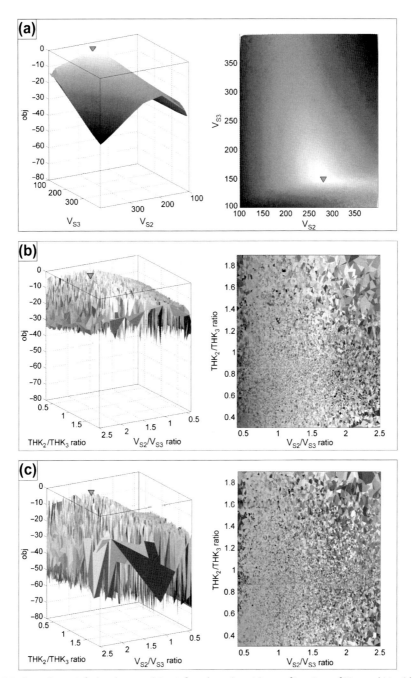

Figure 5.5 Complex misfit landscape: (a) misfit values (z-axis) as a function of V_{S2} and V_{S3} (the other parameters are fixed to their right values) for a 5-layer mode; (b) misfit value as a function of the V_{S2}/V_{S3} and THK_2/THK_3 ratios (the other parameters are fixed to their right values) for a 5-layer model while considering the fundamental mode only; or (c) both fundamental and first higher mode. Triangles indicate the position of the real (reference) model. See text for details. *After Dal Moro and Ferigo (2011).*

dispersion curves, lack of higher modes useful to better constrain the solution etc.—not to mention the problem of the proper interpretation of the velocity spectra in terms of modal dispersion curves), we can actually speak about *equivalent models* (i.e., different models have absolutely similar dispersion curves).

How can the local-minima problem be handled? A traditional approach is represented by the so-called Occam's razor (e.g., Constable et al., 1987): between two or more equivalent models (having the same *misfit* value) choose the *simplest* one (i.e., the one having the smallest variations of the key parameters).

Although this perspective long dominated the way inversion procedures were designed, if we observe the complex meandering of a river or the baroque and articulated shapes of some tectonic structures, it is quite clear that the Occam's principle ("*Pluralitas non est ponenda sine necessitate*") has little relevance when dealing with the unpredictable richness of mother nature (remember the Hamlet's quote opening the previous chapter).

Things can significantly change (improving) if, instead of using the inversion of the modal dispersion curves of a single component, we consider unconventional inversion methods such for instance those introduced in the Chapter 6 and/or the joint inversion introduced in the next Paragraph.

We must surely recall that nonuniqueness is an issue for any methodology that relies on data collected from the surface, not only for surface waves and HVSR. Figure 5.6 reports three subsurface models that, with respect to refraction travel times, result equivalent and thus indistinguishable (Ivanov et al., 2005a; 2005b; 2006). Similarly, Figure 5.7 shows two different density models that produce the same gravitational effect being consequently impossible to differentiate if no other information or dataset is considered (Scales et al., 2001).

5.4 JOINT ANALYSIS

The only way to face all the problems related to the nonuniqueness of the solution and those associated to the interpretation of complex data (in this case fundamentally represented by complex and problematic velocity spectra) is represented by the joint analysis of different and mutually integrating datasets.

Theoretical achievements and performances of the state-of-the-art computers make the joint *analysis* (or, if the reader wishes, *inversion*) one of the most important research topic in several research fields (for an example of applications related to the oil industry—see e.g., Harris et al., 2009).

A synthetic and conceptual representation is schematized in Figure 5.8. In this representation, the method/dataset A (for instance—to be more concrete—the velocity spectrum of the vertical component of Rayleigh waves) can be explained by seven models (A—G) while the method/dataset B by the E—M models. Only some of them (the models G, E, and F) are in common and that means that by considering both the methods/datasets we have now better constrained the solution, by excluding the models

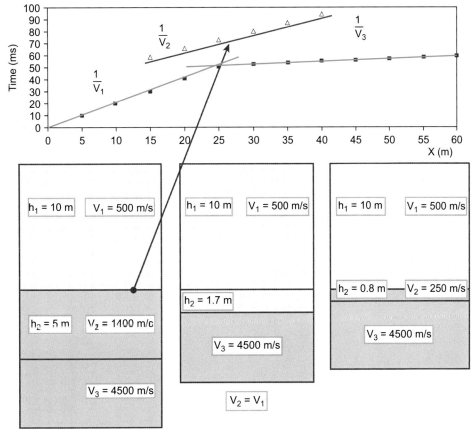

Figure 5.6 Nonuniqueness in traditional refraction seismics: first arrivals can be explained by three totally different models. Please consider that, in case of field datasets, the red refracted event produced by the first refractor in the model 1 on the left are not observed because of the interference with other waves such as surface and guided waves and reflections. *From Ivanov et al. (2006).*

A—D (possible if we would use only the first method/dataset) and the models H—M (that could be used to justify the second method/dataset). This concept can be continued to include more and more objective functions thus constantly reducing the ambiguities that would otherwise taint and jeopardize any inversion procedure based on a single method/dataset (i.e., objective function). In the conceptual example of Figure 5.8, the joint analysis of the three considered components allows to identify the model F as the only one capable of explaining for all the three observations.

In principle, performing a joint analysis is not too different that going through a single-objective function inversion: we can proceed with a direct modeling or with some automatic inversion procedure. Of course, in the second case, we need to properly define the cost function, i.e., defining the way misfit is computed and treated.

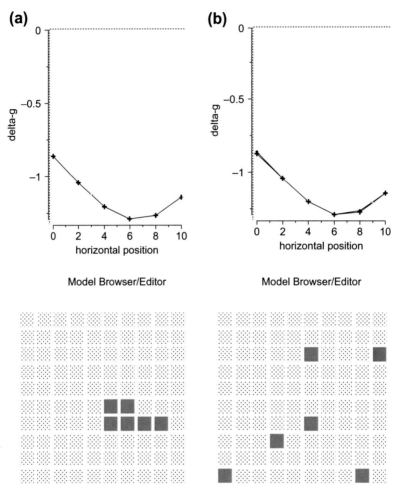

Figure 5.7 Non-uniqueness of the solution in gravimetry: left (a) and right (b) panels report the density distribution (the two gray tones represent two materials with different densities) and the gravimetric effect for two different models characterized by a different density distribution (lower panel). The gravitational effect of both the models is practically the same. *From Scales et al. (2001).*

The classical way to do it is simply represented by the summation of the *misfits* of each single considered objective function (Picozzi and Albarello, 2007). If, for instance, we are dealing with Rayleigh- and Love-wave dispersion, we can define the misfits of the two field data with respect to a tentative model, sum them up, and then seek for a minimum of such a combined cost function.

By following this procedure, a series of problems will anyway inevitably appear. First of all the two (or more) objective functions might be incommensurable and their values might require some normalization that can result tricky and problematic. Furthermore,

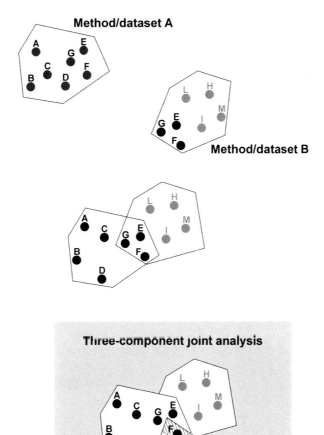

Figure 5.8 Conceptual scheme representing the importance of the joint analysis for reducing the ambiguity and nonuniqueness of the solution: only by using more datasets it is possible to better constrain the solution and identify the *correct* model.

the considered objective functions can result (at least to some degree) conflicting and the resulting combined value can consequently assume an unstable and almost meaningless trend (Fonseca and Fleming, 1993; Van Veldhuizen and Lamont, 1998a,b, 2000; Zitzler and Thiele, 1999; Coello Coello, 2002, 2003).

A possible solution to these problems is represented by the possibility of keeping the single misfits separate and evaluating their values on the basis of a ranking procedure based on the Pareto dominance criterion (a sort of search for the best trade-off among all the considered objective functions/misfits). This sort of approach (the multiobjective optimization described in the cited articles) is actually quite common in economy and in all those fields where a solution representing the best model (or compromise) of a complex (multiobjective) problem is sought (e.g., Yapo et al., 1998). Basically the objective functions are defined as in the single-objective case and the ranking of the models is then computed on the basis of the Pareto criterion while the optimization engine is typically represented by some heuristic (*global search*) schemes such as the Genetic or Evolutionary Algorithms (the resulting full acronym is then consequently typically MOEA—*Multi-Objective Evolutionary Algorithm*).

The application of this approach in the joint analysis of seismic data is presented in Dal Moro and Pipan (2007), Dal Moro (2008, 2010), and Dal Moro and Ferigo (2011) and the fundamental principles are briefly illustrated in the "MOEA in short" box, while a concrete example is presented in the case studies 2 and 4 and in the Paragraph 7.2.

MOEA in Short

One of the key aspects that characterize the multiobjective approach (i.e., the approach that—in a joint inversion—allows us to keep separate the misfit values of the two (or more) considered objective functions) is that its outcomes somehow help us in understanding whether the performed inversion procedures is overall consistent.

Figure 5.9 shows an example of MOEA outcome obtained while considering two objective functions (such, for instance, Rayleigh- and Love-wave dispersion). Each point represents a model and the two associated values (obj 1 and obj 2) are the associated misfit values with respect to the first and the second considered objective functions: for a given tentative model the misfit

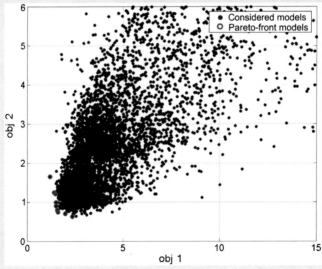

Figure 5.9 Bi-objective space with a cloud of models showing a general congruency of the inversion process. *From Dal Moro and Pipan (2007).*

are computed both with respect to the Rayleigh and Love waves and the two values are used to rank the model by means of the Pareto criterion and proceed with the optimization.

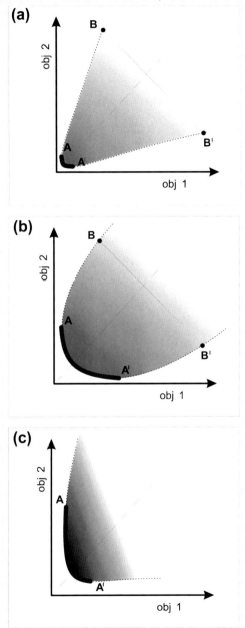

Figure 5.10 Criteria for interpreting the model distribution in a multiobjective space (in this case considering just two components): (a) highly consistent inversion process; (b) slightly conflicting objective functions (minor problems); (c) highly conflicting objective functions (inconsistent inversion procedure). *After Dal Moro and Pipan (2007).*

(Continued)

In the end all the evaluated models are shown in the objective space and the distribution of the models will provide a hint of consistency of the overall accomplished inversion procedure.

Ideally the misfit values for the two (or more) objective functions should approach as close as possible to the [0, 0] point which, not by chance, is therefore usually indicated as the *utopia point*.

If the inversion procedure was properly managed, the cloud of models should clearly point toward the *utopia point*, thus showing to converge in a consistent way (see Figure 5.9).

Anyway, if we decide to invert the data following the common *modal* dispersion-curve picking, some problem with the data interpretation can occur (erroneous mode identification etc.—see Chapter 3). In that case the joint inversion will necessarily face some problem that will necessarily mirror in an *anomalous* distribution of the models in the objective space.

In that case the symmetry of the Pareto models with respect to the *universe* of evaluated models will diverge from the symmetric trend shown in Figures 5.9 and 5.10(a) by an amount that will depend on the seriousness of the problem itself.

An erroneous mode identification will typically produce large deviations from the symmetry (Figure 5.10(c) and case study 4) while smaller inconsistencies (such as those, for instance, created by V_{SH}–V_{SV} anisotropies or minor interpretative issues) will create a less pointy cloud of model such as the one shown in Figure 5.10(b).

It goes without saying that this approach is actually viable while considering both the traditional dispersion-curve picking and the unconventional methods introduced in the next chapter.

CHAPTER 6

Full Velocity Spectrum Inversion and Other Unconventional Approaches

Act from principle.
Move from intention.

Robert Fripp

As shown in the previous chapters, because of their constitutive equations, Rayleigh waves can propagate following a complex phenomenology thus resulting in nontrivial mode excitement that mirrors in complex velocity spectra.

It was also shown that Love-wave velocity spectra are typically by far simpler and consequently easier to interpret in terms of modal dispersion curves (a series of case studies putting in evidence this fact is presented in the Appendix) but, on the other hand, are slightly so-to-say less sensitive to small low-velocity channels (V_S inversions).

These aspects can be faced by adopting so-to-speak unconventional approaches capable of solving possible interpretative issues also reducing the nonuniqueness (i.e., the intrinsic ambiguity) of the solution.

This chapter will present some approaches not based on the identification and inversion of the modal dispersion curves, which, when data are not following simple trends, can result quite hard to identify and understand.

The concepts of *effective* dispersion curve, *full waveform* and *full velocity spectrum* inversions will be briefly presented to evidence pros and cons of their application.

6.1 INTRODUCTION

In order to synthetically recap the problems that cannot be easily faced and solved through the analysis of the modal dispersion curves (see Chapter 3), we will consider the model reported in Figure 6.1(a) and that can be thought as representative of a situation of a 2.5 m soft-sediment cover over a gravel-like layer (with a total thickness of about 10 m with deeper part slightly "weaker" than the first 5 m). A hard bedrock is eventually present at a depth of about 14 m.

Synthetic traces pertaining to the ZVF component and computed according to Carcione (1992) are shown in Figure 6.1(b) while the phase-velocity spectrum is presented in Figure 6.1(c) and shows a remarkable fact that requires some comments.

Although the theoretical modal dispersion curves of the first four modes (of the considered model) are plotted over the velocity spectrum, the reader should consider

Surface Wave Analysis for Near Surface Applications
ISBN 978-0-12-800770-9, http://dx.doi.org/10.1016/B978-0-12-800770-9.00006-6

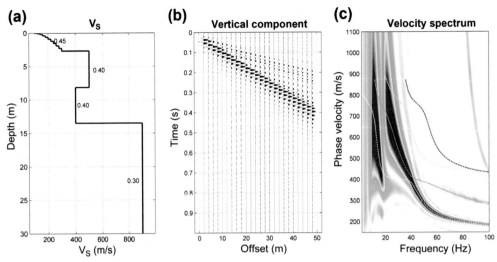

Figure 6.1 A synthetic dataset: vertical component of Rayleigh waves: (a) V_S model (reported numbers are the adopted Poisson ratios), (b) synthetic traces, and (c) velocity spectrum with, overlain, the modal dispersion curves.

that while dealing with field datasets the V_S model is unknown and the proper under-standing of the velocity spectrum is the key point to face.

Let us then imagine that the dataset comes from a field acquisition and consequently modal dispersion curves plotted over the velocity spectrum are not available and our task is the understanding of the velocity spectrum. The most striking evidence in this case (see Figure 6.1(c)) is the presence of a "jump" at about 20 Hz while the signal before and after this point appears continuous.

Having at our disposal only these data and considering the continuity of the two "branches" before and after the jump at 20 Hz, the interpretation would likely, easily, and necessarily involve the fundamental mode as responsible for the low-frequency part (frequency lower than 20 Hz) and the first higher mode as responsible for the disper-sion occurring at higher frequencies.

Unfortunately this sort of interpretation would be wrong since, if we now carefully examine the theoretical modal dispersion curves plotted over the velocity spectrum, we finally see that the continuous signal for frequencies higher than 20 Hz is actually the combination of two different modes: the first overtone between 20 and 40 Hz and the fundamental one for frequencies higher than 40 Hz.

What is quite striking in this simple but illuminating example is that the continuity of a signal does not necessarily mean that a single mode is responsible for it (a fact unfortu-nately often poorly understood and rarely considered).

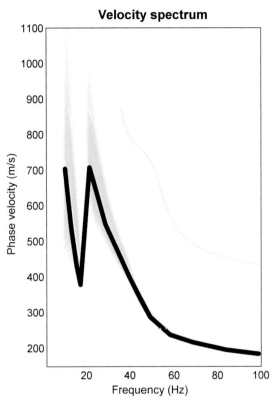

Figure 6.2 Effective dispersion curve of the dataset presented in Figure 6.1. The thick black curve represents the effective signal the practitioner will have to deal with during the analysis (see text for explanation).

For the sake of clarity, in Figure 6.2 is reported the so-called *effective* dispersion curve for the considered (synthetic) dataset.

In short, it must be kept in mind that sometimes the way surface-wave energy unfolds does not follow simple trends and, consequently, velocity spectra (in particular those related to Rayleigh waves) cannot always be interpreted in terms of simple modal dispersion curves, since the effective dispersion curve can actually refer to complex (somehow even counterintuitive) mixture of modal curves (the relation between modal and effective dispersion curves is described for instance in Tokimatsu et al., 1992).

There are two possible ways to face and possibly solve this problem: using a joint approach capable of addressing this sort of ambiguities or following an approach not simply based on modal dispersion curves but on some more sophisticated computation (i.e., more complete mathematical approach and solution). In addition to the implementation of the effective dispersion curve, two further possible solutions will be briefly introduced in the following.

Actually, at least to some degree, the effective dispersion curve is offset dependent (please consider that the velocity spectra obtained from field datasets while considering multichannel – i.e., multioffset – acquisitions clearly represents a kind of average velocity spectrum resulting from the combination of the effective dispersion curves along the array).

Figure 6.3 reports the effective dispersion curves (ZVF component) computed for a number of offsets (for the reported V_S model) according to Lai and Rix (1998; 2002). In

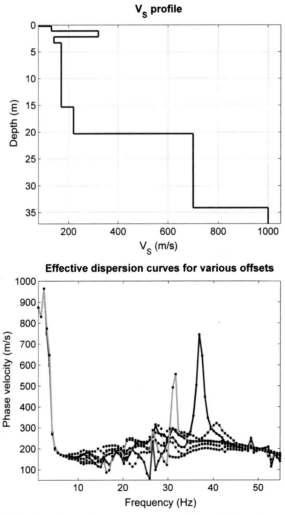

Figure 6.3 A series of effective dispersion curves computed considering the model reported in the upper panel while considering different offsets (3–65 m).

this case, the way-around to adopt to invert the field data can be easily represented by the average effective dispersion curve for all the considered offsets.

While for an overview on these topics can be found for instance in Tokimatsu et al. (1992) and Foti et al. (2000), further approaches could also be considered and are briefly presented in the next Paragraph.

6.2 FULL WAVEFORM AND FULL VELOCITY SPECTRUM INVERSIONS

The *Full Velocity Spectrum* (FVS) procedure can be somehow regarded as a sort of extended *effective* dispersion curve approach which, computationally speaking, result less intensive with respect to the *Full Waveform Inversion* (FWI) (Forbriger, 2003a,b; Groos, 2013; Groos et al., 2013) which, dealing with the representation of the entire wavefield, requires several hours of computation times on current state-of-the-art supercomputers. In this latter approach (FWI), quite attractive in a number of applications where a detailed solution is relevant such as, for instance, the oil industry, the optimization is performed through the minimization of the difference (the misfit) between the observed and the computed seismic traces, thus basically working in the time domain and, in order to be properly performed, requires the knowledge or the estimation of the source wavelet (Figure 6.4).

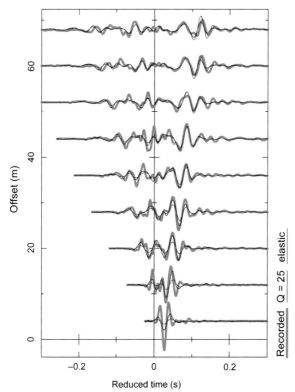

Figure 6.4 Comparison of vertical particle velocity seismograms of field data (thick greenish line) and synthetic data after application of a source wavelet correction and while considering the elastic and viscoelastic cases. *From Groos (2013).*

In order to deal with an approach capable of solving the problems briefly recalled in the introductory paragraph but dealing with a less-intensive computational load, a different approach can be followed.

In fact, since here we are interested only in the low-frequency surface waves (and not in the entire waveform which would also include higher-frequency reflected, refracted, and guided waves), the computation of the synthetic traces can be efficiently performed via modal summation (e.g., Herrmann, 2003).

Moreover, since the dispersive properties basically depend on the medium and not on the specific wavelet that characterizes the seismic source, if we work in the frequency—velocity domain we can set up an inversion procedure that can be used to invert surface-wave velocity spectra while ignoring the source wavelet (a schematic representation of this approach is reported in Figure 6.5) (see also Dal Moro et al., 2014).

Quite clearly, all these approaches can be adopted while considering both single- and multicomponent data (joint analysis to perform according to the different possible schemes mentioned in the previous chapter).

Here the fundamental "object" considered during the inversion procedure is not any longer a picked dispersion curve as in the traditional approach nor the acquired seismic traces (as in the FWI approach) but the whole velocity spectrum matrix (i.e., the matrix representing the data correlation as a function of frequency and velocity) being the final goal the identification of a subsurface model that properly describes, i.e., mimics, the velocity spectrum of the field dataset.

In Figure 6.6 is presented an example of the sought outcome (in this case data were collected on a sedimentary cover overlaying a bedrock at about 20 m of depth): the

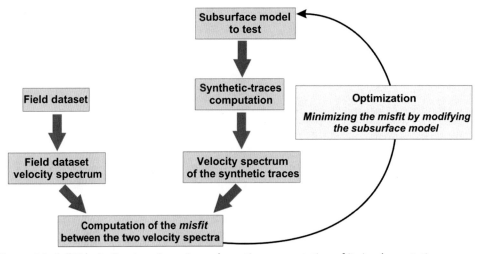

Figure 6.5 *Full Velocity Spectrum* inversion: schematic representation of its implementation.

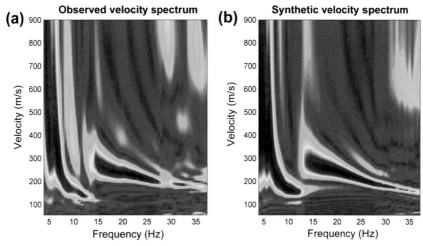

Figure 6.6 Example of *Full Velocity Spectrum* inversion: (a) velocity spectrum of the field data, (b) velocity spectrum of the identified model (i.e., the one minimizing the velocity spectrum misfit as defined in the scheme reported in Figure 6.5).

phase-velocity spectrum of the synthetic model (b) closely "resembles" the observed one (a). Figure 6.7 reports the same data while adopting a more synthetic representation (superimposition of synthetic and observed velocity spectra).

Also keeping in mind all the aspects reported in the introductory paragraph, it must be underlined that the signal dominating the velocity spectrum for frequencies higher than 15 Hz could be in principle related not to a single mode but, following this approach, no interpretation is actually made (exactly the same way as we do while considering the approach based on the *effective* dispersion curve). A certain degree of ambiguity is necessarily still implicit also in this approach and only the joint analysis with further components can eventually definitely *solve the site*.

Some case studies solved by adopting the FVS approach are reported in the Appendix, both in the framework of single- and multicomponent (i.e., joint) inversions (see, in particular, case study 7 where an FVS joint analysis of the RVF and THF components is presented).

Compared to the classical approach based on modal dispersion curves, the FVS approach is surely computationally heavier and, in order to obtain good performances in terms of results and computational load, requires a careful choice of a series of parameters. For instance, the velocity spectrum to invert must be focused on the relevant part of the signal. In Figure 6.6 the spectrum was fixed in the 4–37 Hz frequency range for phase velocities between 50 and 900 m/s, but these parameters essentially depend on the site, the acquisition geometry, and the goal of the survey.

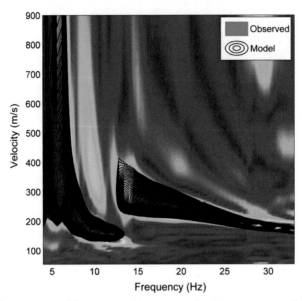

Figure 6.7 Same data presented in Figure 6.6 but now adopting a more synthetic representation: background colors report the phase-velocity spectrum of the field data while overlaying black contour lines the one of the identified model.

Since through our synthetics we are simulating only surface waves, data should also be cleaned up from possible large-amplitude refracted events or guided waves (see case study 2). Moreover, before computing the velocity spectrum of the field dataset, it is also good practice to remove from the data all the useless time samples. If, for instance, we consider the seismic traces in Figure 6.1(b), in case we intend to go through an FVS inversion, we should keep the seismic traces only down to 0.6 s (removing the rest).

Finally, the parameters that regulate the generation of the synthetic traces must be carefully chosen. If, for instance, we intend to use the above-mentioned modal summation, it is important to properly fix the number of modes to use: an insufficient number of modes would clearly result in an incapability of properly representing the observed data (in case the dataset is characterized by a significant amount of energy in the higher overtones), while a uselessly large amount of modes would mirror in longer computational times with no gain in terms of goodness of the solution.

Needless to say that the effects of attenuation should also be considered (see, e.g., Panza, 1989 and Chapter 3). Figure 6.8 reports the results of the inversion performed for the same data considered in Figure 6.6 and 6.8 but considering the approach based on the *effective* dispersion curve (i.e., without any interpretation of the velocity spectrum in terms of modes). In this case the two approaches are absolutely equivalent and provide the same results, but, in some special cases (see next Chapter and some of the reported case studies), the specific characteristics of the data may require to follow one approach

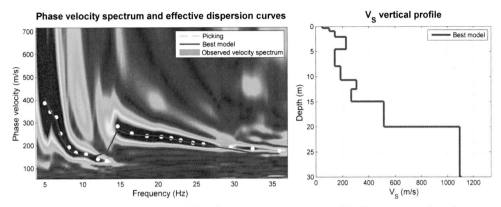

Figure 6.8 Same dataset considered for the FVS analysis presented in Figure 6.6 and 6.7 but now solved while considering the picking and inversion of the *effective* dispersion curve (no interpretation in terms of modal curves).

rather than the other (needless to say that the FVS analysis is computationally heavier than the one based on the *effective* dispersion curves).

In any case, independently on the approach we intend to adopt to automatically invert the data, some preliminary forward modeling (see previous chapter) result extremely useful to understand the best parameters to adopt and have a reasonable starting model to optimize.

CHAPTER 7

Some Final Notes

Performing is very much like cooking: putting it all together, raising the temperature

David Tudor

7.1 THE ADOPTED PERSPECTIVE

The slant and the architecture of the entire book were designed while considering an austere motto: *if you don't understand everything, you won't understand anything.* Not by chance, a number of cross-references are in fact present all over the book: the practical consequences of some theoretical facts are shown by referring to the data presented in some case studies and, vice versa, some analyses presented in some case studies are explained and corroborated by referring to some theoretical evidence presented in the previous chapters.

This necessarily reflects in a continuous back and forth reading but, once all the relevant aspects are correctly acquired, it is possible to safely proceed with adequate acquisitions and analyses.

The leitmotif of the book (so the Tudor's quote and the "everything or nothing" motto) is in fact that the so-to-speak holistic approach (i.e., the joint analysis of multiple components) is the sole solution to the ambiguities and the nonuniqueness intrinsically present in the data and, thus, in the analyses.

As illustrated in Chapter 5, seismic data inversion must be performed taking care of a number of possible issues. Actually, there is no single and universal approach and, for obvious economical reasons, the *best* approach (i.e., the one capable of properly and unambiguously solving the site while minimizing the costs) must be tailored on the specific site and considering the specific goals to achieve. Some stratigraphic conditions produce very complex data that necessarily require the collection of multicomponent and multichannel datasets (see, e.g., case study 11) while, in other cases, a very simple single-geophone acquisition campaign can be sufficient (see group-velocity analyses presented for the case studies 2, 6 and 8 and later on in this chapter).

In any case, analyzing the data means first of all understanding them: the link between the appearance of a signal in the *time—offset* domain (the acquired seismic traces) and its features in the *frequency—velocity* domain (velocity spectra) must be clear. Any *black-box* procedure that promises to provide the *right* solution without a serious understanding of the data by the side of the practitioners is necessarily highly risky, especially if based on a single component whose solution cannot be cross-checked and validated by further components (see the very basic concept of joint analysis presented in Chapter 5).

Surface Wave Analysis for Near Surface Applications
ISBN 978-0-12-800770-9, http://dx.doi.org/10.1016/B978-0-12-800770-9.00007-8

This is a quite relevant issue because nowadays surface-wave analysis is not a mere academic research field but also a tool widely used in a number of geotechnical studies, so with a significant impact on the society.

Buying the necessary equipment and software does not ensure anything because data analysis is very much like the Tudor's performance: temperature (which in this as in other cases represents the *knowledge* of all the relevant facts) must be raised.

7.2 A BRIEF MISCELLANEA ON MODES AND SHEAR-WAVE VELOCITIES

In this paragraph, we will become aware of some further ancillary facts related to surface-wave propagation and therefore useful while analyzing the data.

Understanding the modes from a practical point of view is probably the biggest problem that practitioners must face. The whole Chapter 3 was devoted to the description of surface-wave phenomenology which, fundamentally, is right about the understanding of how different modes can interlace and result more or less prominent in a given dataset. In spite of the name, the *fundamental* mode does not have to be considered more important than others. Actually, because of the higher phase velocities, higher modes bring relevant information especially for the deepest layers and must be consequently always properly considered clearly keeping in mind that, often, the apparent dispersion curve cannot be simplistically described in terms of single modal curves (see, e.g., case study 11).

The apparent dispersion curve is in fact the result of the superposition of all the modes (from the point of view of the velocities), and its trend depends surely on the site but also on the source and considered component. Incidentally, horizontal-to-vertical spectral ratio (HVSR) curve is the result of the same things but from the point of view of the (relative) amplitudes.

7.2.1 More about Modes

Data reported in Figures 7.1 and 7.2 put in evidence a couple of facts. A field dataset (in this case the radial component of Rayleigh waves—RVF) was filtered in the *f-k* domain (see also case study 9; Yilmaz, 1987; Baker, 1999; Dal Moro, 2011) to put in evidence some specific modes in the 20–60 Hz frequency range. The large signal (large correlation values) for frequencies smaller than 20 Hz is clearly related to a coalescence of various modes that cannot be singularly resolved and is therefore entirely kept.

A first filter is applied (Figure 7.1) to remove the contribution of higher modes for frequencies higher than 20 Hz (but it must be considered that their contribution for lower frequencies is still there) and the result is that the obtained seismic traces emphasize the "late" arrivals related to the low-order modes (travelling slower than the higher modes—see dashed blue polygon).

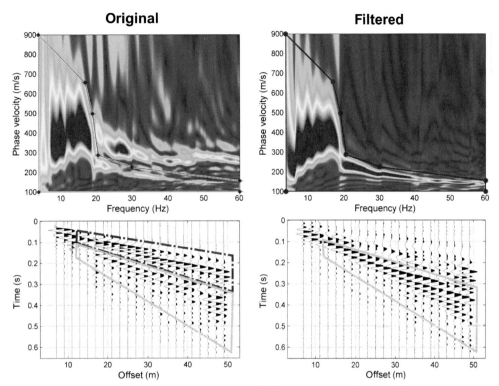

Figure 7.1 Separating the mode contribution via *f-k* filtering: on the left panel, the original dataset (seismic traces and phase-velocity spectrum); on the right, the data filtered in the *f-k* domain while considering the indicated red polygon (kept the data within the polygon). The low-order modes for frequencies higher than 20 Hz are now emphasized (see seismic traces highlighted by the greenish polygon in the *offset–time* domain).

On the other hand (Figure 7.2), if we now keep only the contribution of higher modes at frequencies higher than 20 Hz, we emphasize the early arrivals (dashed red polygon) related to the (faster) higher modes. Incidentally, since we removed all the frequencies lower than about 20 Hz, for those frequencies the resulting phase-velocity spectrum assumes that peculiar (fundamentally meaningless) aspect.

In this case, for frequencies below 20 Hz, the wide range of exited phase velocities clearly suggests that a number of modes (which cannot be singularly resolved) are involved. The consequence is that, in this case, an approach based on the apparent dispersion curve (that will be used later on for inverting another dataset) would be problematic, while the Full Velocity Spectrum (FVS) inversion that considers the velocity spectrum as a whole (see Chapter 6) could be more appropriate. Results are presented in Figure 7.3 and show again how complex can be the way modes excite and interlace. It is in fact, for instance, not very intuitive that the A signal highlighted

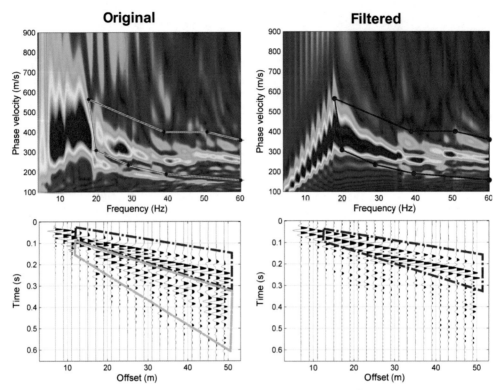

Figure 7.2 Separating the mode contribution via *f-k* filtering: on the left panel, the original dataset (seismic traces and phase-velocity spectrum); on the right, the data filtered in the *f-k* domain while considering the indicated red polygon (kept the data within the polygon). The higher-order modes for frequencies higher than 20 Hz are now emphasized (see seismic traces highlighted by the red polygon in the *offset–time* domain).

in Figure 7.3(d) refers not to the first higher mode (as probably simplistic interpretations would suggest) but to the third higher mode (compare with Figure 7.3(f)), while for frequencies lower than 20 Hz, the gross of the energy pertains to the first and second higher modes.

It is essential to underline again that the FVS approach is fundamentally different than the Full Waveform Inversion since the synthetic traces are computed via modal summation and consequently the computational time is significantly lower than the one necessary in the FWI approach. Since, unless of detailed analyses, the actual source wavelet is unknown and only surface waves are considered (so refractions, reflections, and guided waves are not included) while adopting the FVS approach, it is consequently impossible to seek for a perfect match of the traces in the *offset–time* domain; although, considered the amplitude of the surface waves, the general trend must be similar (compare the good general agreement of field and synthetic traces in Figure 7.3(a) and (c)).

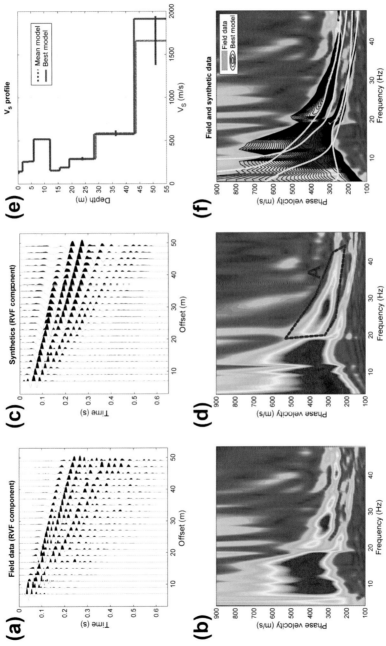

Figure 7.3 *Full Velocity Spectrum* (FVS) inversion of the dataset (RVF component) considered for the *f-k* filtering (mode separation) presented in Figures 7.1 and 7.2: (a) and (b) present the field seismic traces and the respective phase-velocity spectra, (c) and (d) report the synthetic seismic traces and the respective velocity spectra for the model identified via FVS invers on, (e) the best (minimum misfit) and the mean models (see Dal Moro et al., 2007), (f) background colors report the observed velocity spectrum (same as in plot (b)), overlaying contour lines the synthetic velocity spectrum (same as in plot (d)) while the yellow lines represent the modal dispersion curve of the identified model. See text for comments.

7.2.2 Water

Does water in porous media influence shear-wave velocities?

In the case studies 2, 5, 8, and 9, the P-wave refraction related to the water table is presented and considered to characterize the site but, in general terms, the topic can be surely faced theoretically and via lab measurements (e.g., Prasad, 2002).

Considering the behavior of very shallow porous unconsolidated sediments, while the water strongly increases V_P values, the effect on shear-wave velocities can change depending on the saturation conditions. At low saturation levels, an increase in the moisture tends to produce higher shear-wave velocities while, in fully saturated media, pore fluid pressure decreases the shear velocity (West and Menke, 2000).

A quick and simple field experiment (the acquisition and analysis of the group velocities determined while using only a single 3-component (3C) geophone) can be easily performed on a sandy beach.

Figure 7.4 shows the experimental setting: two acquisitions were performed on the inland dry sands of a beach (Line A) and on the very shoreline (Line B) where water saturates the very shallow layers. The two lines are approximately 50 m one from the other.

Data from the 3C (calibrated) geophone were analyzed following the same joint procedure used, for instance, for the case studies 2 and 6 (a multiobjective evolutionary algorithm inversion scheme handling the three considered objective functions: group-velocity spectra of the ZVF and RVF components jointly with the *radial-to-vertical spectral ratio* (RVSR)).

Results of the analyses performed on the two datasets are reported in Figure 7.5 for Line A (the dry sands of the inland area) and Figure 7.6 for Line B (the saturated sands the shoreline).

Actually, the difference in V_R (Rayleigh-wave group velocities) (so in the V_S as well) is quite apparent even simply giving a look at the group-velocity spectra of the field data reported (the background colors) in Figures 7.5 and 7.6: for frequencies higher than about 5 Hz, the group velocities along the Line A is about 180 m/s, while along the Line B (on the shoreline) the V_R group velocity drops to more or less 140 m/s.

Shear-wave velocities identified for the shallowest layers (around 200 m/s for the dry sands and around 160 m/s for the saturated sands of the shoreline) fundamentally support the results of West and Menke (2000) that, considering the water table variations induced by the tides in unconsolidated sands (Figure 7.7), indicate a V_S decrease of *at least* 15%. In our case, the V_S decreases by about 20% and, by assuming that the composition and texture of the materials are fundamentally the same, such a decrease can be attributed to the water saturation.

7.2.3 Peats

In terms of geomechanical properties, peats are probably the poorest material on Earth (in the author's experience, only the Lunar regolith—case study 14—beats the terrestrial

Figure 7.4 Effect of water saturation on shallow beach sands: experimental setting. Two active acquisitions were performed along the A and B line while considering a single 3-component geophone (offset 60 m) and by analyzing the group-velocity spectra of the radial and vertical components of Rayleigh waves also jointly with the *radial-to-vertical spectral ratio (see Figure 7.5 and 7.6)*.

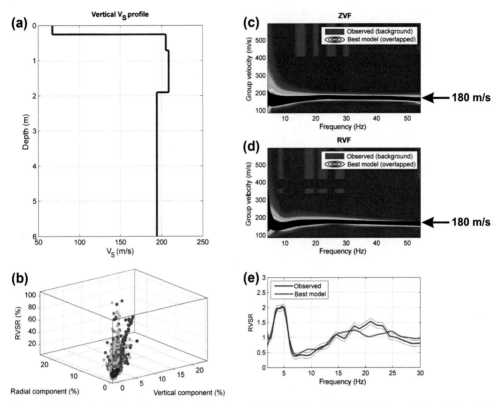

Figure 7.5 Results of the Line A data (inland area): (a) obtained V_S profile down to 6 m, (b) misfit distribution for the three considered objective functions (misfits of the ZVF and RVF group velocity spectra and radial-to-vertical spectral ratio (RVSR)), (c) and (d) observed (background colors) and synthetic (overlying black contour lines) group-velocity spectra for the ZVF and RVF components, (e) observed and synthetic RVSR.

peats). Shear–wave velocities of soils particularly rich in organic matter are notoriously extremely low and the data presented in Figure 7.8 give a straightforward evidence of it. The seismic traces themselves (with no need for sophisticated analyses) can immediately give a clear evidence: the slope of the propagating Rayleigh wave shows that the time necessary to cover the 56 m of the array is (on an average) more than 1 s. Since the average V_R is obtained by simply dividing *space* by *time*, it is instantly clear that we are dealing with velocities around 50 m/s.

Since the presence of a *mode splitting* phenomenon (see Chapter 3) is quite apparent, the inversion procedure was performed while considering the *apparent* dispersion curve, thus not by interpreting the velocity spectrum in terms of modal dispersion curves but properly considering the modal superposition (Tokimatsu et al., 1992). A genetic algorithm inversion scheme (Dal Moro et al., 2007) was adopted with the results reported

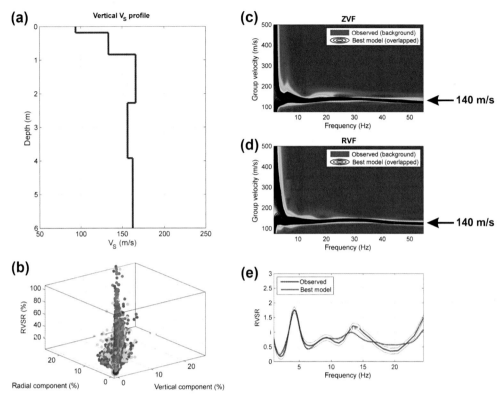

Figure 7.6 Results of the Line B data (shoreline): (a) obtained V_S profile down to 6 m, (b) misfit distribution for the three considered objective functions (misfits of the ZVF and RVF group velocity spectra and radial-to-vertical spectral ratio (RVSR)), (c) and (d) observed (background colors) and synthetic (overlying black contour lines) group-velocity spectra for the ZVF and RVF components, (e) observed and synthetic RVSR.

in Figure 7.8 and giving evidence of the peat level, characterizing the sequence down to a depth of about 6 m.

7.3 SURVEY PLANNING AND RESULT EVALUATION

A careful planning of the survey is essential when the objective is to provide a reliable and robust V_S model and the presentation of the results should always give the chance to evaluate the quality of the accomplished work.

Fundamentally, we can consider two different kinds of surveys. Sometimes it is in fact required the V_S profile along a single vertical profile (very often this is done in the framework of seismic-hazard analyses or for geotechnical studies related to the construction of small buildings) while in other cases (when the goal is the exploration of a more or less wider area) 2D acquisitions and multiple shots must be considered.

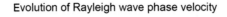

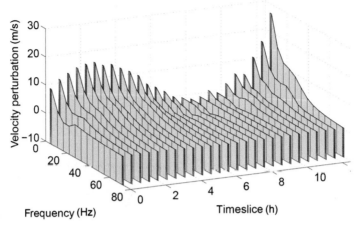

Figure 7.7 Changes in Rayleigh-wave velocities induced by variations of water saturation in uncon-solidated sands: reported the velocity perturbation as a function of frequency and elapsed time (from a reference zero time). The largest variations in velocity occur in the low frequencies sensing the variations in the water table depth induced by the tides. *From West and Menke (2000).*

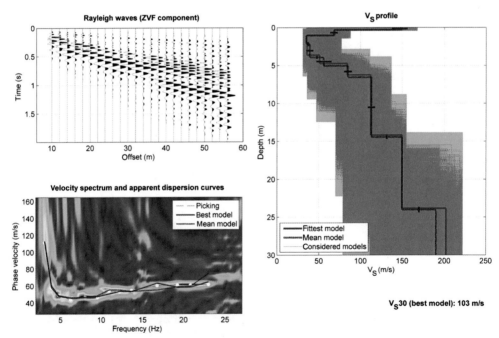

Figure 7.8 Dataset (ZVF component) acquired in an area where a peat level is present between about 1 and 6 m of depth. The genetic algorithm (GA)-based inversion was performed while considering the *apparent* dispersion curve computed according to Tokimatsu et al. (1992). In the plot reported on the right, the green area represents the adopted search space while the gray lines represent the totality of the models considered during the GA-based inversion procedure.

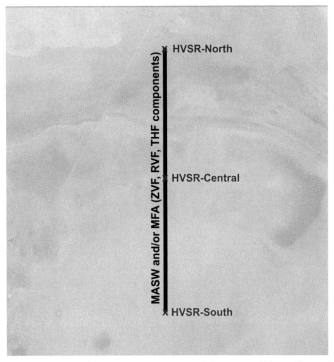

Figure 7.9 A classical acquisition for the reconstruction of the vertical V_S profile: active (multicomponent) and passive (Horizontal-to-Vertical Spectral Ratio, HVSR) data can be considered to then proceed with the joint inversion capable of solving nonuniqueness of the solution and various possible ambiguities necessarily present in each single dataset. MASW, multichannel analysis of surface waves; MFA, multiple filter analysis.

It is clear that when a single vertical profile is required, the field procedures can be planned in such a way to easily acquire multicomponent data useful for accurate analyses. Figure 7.9 reports the simplest case: surface-wave dispersion is considered along a line according to multichannel (phase-velocity Multichannel Analysis of Surface Waves - MASW) or single-geophone (group-velocity Multiple Filter Analysis - MFA) acquisitions. In both cases, it is highly recommended to acquire multicomponent data.

If we can assume no major lateral variations, then we can acquire the passive data for the computation of the H/V spectral ratio only in the central point, while if we intend to verify if major lateral variations are present, we can record further H/V data at the ends of the line (see HVSR South and North in Figure 7.9).

We can also decide to analyze surface-wave dispersion according to passive techniques such as Extended Spatial Autocorrelation Method (ESAC), SPAC (SPatial Auto-Correlation), Refraction Microtremor, and so forth. Figure 7.10 reports an example of acquisition scheme where a bidimensional array is used for ESAC analyses.

In that case, the apparent dispersion curve that will be obtained will depend on the average stratigraphy within the area defined by the three vertices and, consequently, the microtremor acquisition for the determination of the H/V spectral ratio should be performed more or less in the barycenter.

Needless to say that, by exploiting such geometry, it is also possible to acquire two perpendicular active acquisitions that, in Figure 7.10, are indicated as MASW 1 and MASW 2 (reader should always recall that—as widely illustrated throughout this book—by MASW we do not necessarily/only mean vertical geophones and vertical source).

In the final report, the model eventually provided should be always accompanied by all the data necessary to understand and evaluate the accomplished analyses. Throughout the whole book, it was in fact shown how difficult can be the identification of the modal dispersion curves (especially when following an approach based on the modal dispersion curves of Rayleigh waves) and that, as a matter of facts, data misinterpretation is consequently quite common.

Providing only picked curves and obtained V_S profile (Figure 7.11(a)) is therefore totally insufficient because, this sort of arrangement does not give the reader the chance to

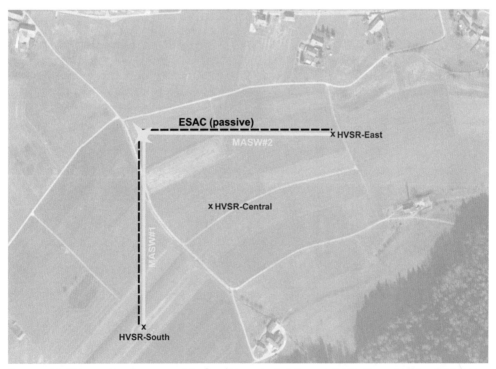

Figure 7.10 Bidimensional array used for acquiring passive data useful for Extended Spatial Autocorrelation Method (ESAC) analysis as well as for two perpendicular multichannel analysis of surface waves (MASW) lines (MASW 1 and MASW 2).

understand how complex was the dataset and what was the criterion followed during the picking/analysis.

On the other side, a more correct and complete way of presenting the analyses is shown in the lower panel (Figure 7.11(b)) where is in fact shown also the velocity spectrum. Incidentally, the two V_S profiles shown in Figure 7.11(a) and (b) are a very simple example of nonuniqueness of the solution: the dispersion curves are exactly the same but below a depth of about 10 m, the two V_S profiles are clearly different.

Of course, considered all the arguments presented in the previous chapters and in the following case studies, it should be clear that, compared with the others, the approaches

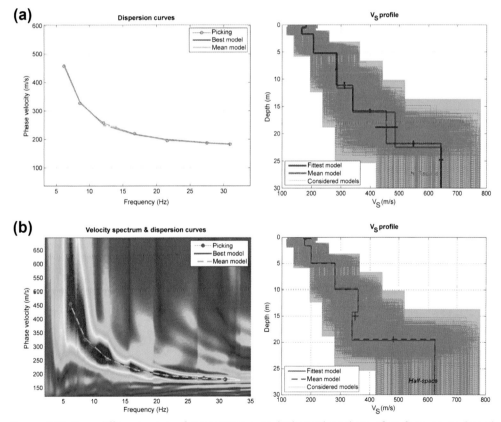

Figure 7.11 Two different ways of presenting a multichannel analysis of surface waves based on modal dispersion curve picking. In the upper panel (a), only the picked and inverted curves are shown while in the lower panel (b) is presented also the velocity spectrum. Only the completeness of this second presentation allows the evaluation of the performed analyses which, in this case, are based on the interpretation of the velocity spectrum in terms of modal dispersion curve(s).

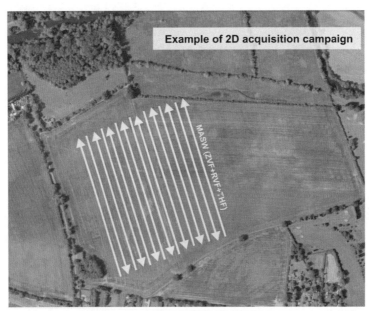

Figure 7.12 Planning the acquisition necessary to explore a vast area via multi-component MASW (Multichannel Analysis of Surface Waves).

based on the modal dispersion curve(s) are often quite problematic and surely the less comprehensive.

When the goal is the exploration of a more or less vast area, things can become quite complex because the acquisition procedures and the processing require special care. First of all, since planting and unplanting the geophones would be practically unfeasible, a landstreamer becomes necessary and a series of back and forth 2D lines must be planned (Figure 7.12). As always, adequate multicomponent acquisitions remain highly desirable, if not mandatory.

Passive seismics becomes impossible to perform on a systematic basis but a few HVSR datasets could be anyway considered in some critical locations of the area to investigate.

Usually this sort of surveys are performed with the aim of identifying lateral variations, but in order to accomplish such a meticulous objective, instead of considering the standard *common shot gathers* (CSGs), data should be rearranged according to the *common midpoint* (CMP) gathers (see Figure 7.13).

It was in fact shown (Shtivelman, 1999, 2002; Hayashi and Suzuki, 2004) that the identification of small-scale lateral inhomogeneities highly benefits from considering CMP datasets rather than the raw CMP gathers from the field.

A series of successive CMP gathers are then processed (seeking individual solutions for each of them) and a contouring is eventually used to image the section (or the 3D block).

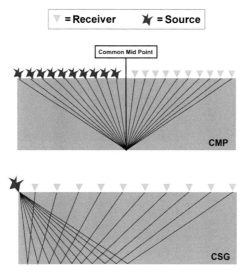

Figure 7.13 Common midpoint (CMP) gathers and common shot gathers (CSGs).

The inversion of such a large amount of data should be performed without neglecting all the illustrated issues (the large amount of data to handle should not reflect in a lowering of the quality), and different strategies could be fixed to efficiently handle the 2D problem. Once a good solution is found for the starting shot, the successive can be, for instance, treated assuming variations only for the velocities (Figure 7.14(a)), only for the thicknesses (Figure 7.14(b)), or, in the most general case (that could reflect in a serious computational effort), for both (Figure 7.14(c)).

7.4 SUMMARIZING FEW FINAL RECOMMENDATIONS

Although guidelines are too often regarded as strict rules to blindly follow independently on any other consideration and we hope that the book somehow helped in having a more comprehensive view, in the following are summarized some basic points to consider while acquiring and analyzing surface wave propagation:

- About active acquisitions for MASW or MFA/FTAN studies. Love-wave velocity spectra are invariably simpler to understand: never forget that when you have to decide what kind of equipment to buy or what kind of acquisition procedures are the most appropriate for the site you are going to explore. If we are interested in multichannel analyses, sometimes it can be quite convenient to buy (only) horizontal-component geophones that allow the acquisition of both Rayleigh (radial component) and Love waves, as well as SH-wave (i.e. the horizontal component of shear waves) refraction/reflection while, in some cases, a single 3C geophone can be profitably used for analyzing group velocities and RVSR (and HVSR as well).

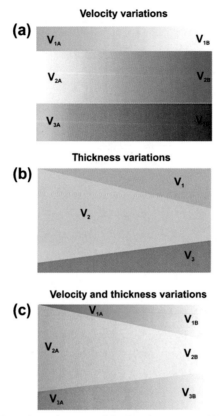

Figure 7.14 Dealing with 2D sections: three possible situations to consider while inverting 2D data for the detection of lateral variations.

- In general terms (for all the active and passive multichannel acquisitions aimed at determining the phase velocities): the larger the array, the bigger the wavelengths that can be properly sampled and, consequently, the lower the frequencies which eventually reflect in a larger penetration depth (a simple rule of thumb suggests that the penetration depth is approximately half of the array length).

- Passive multichannel acquisitions performed while considering bidimensional arrays (eventually analyzed according to the SPAC or ESAC methodologies) are inherently capable of considering the directionality of the signal, whereas passive acquisitions performed while considering a linear array are not.

- MASW, MFA, ESAC, ReMi, SPAC and so forth are different techniques to acquire data useful to depict the dispersive properties but the way these information are then used (inverted) is a completely different story (see Chapters 5 and 6).

- Dispersion analysis: even if the single-component analyses based on the apparent dispersion curve or on the FVS approach are necessarily less prone to misinterpretations and pitfalls, the only way to properly depict the V_S profile is represented by the joint analysis of several components that must be chosen according to both the specific goals and logistical facts. Acquisitions to perform in the framework of detailed 1D V_S reconstruction accomplished for seismic-hazard studies would require Rayleigh- and Love-wave acquisitions both according to active and passive methods (clearly jointly with HVSR measurements) (see, for instance, the acquisition schemes in Figures 7.9 and 7.10); seismic surveys aimed at exploring wider areas (e.g., scheme in Figure 7.12) would almost necessarily rely on purely-active acquisitions and analyses of multicomponent data while considering phase and/or group velocities.

- Properly naming the field data files can be regarded as a trivial fact but, especially when data must be shared with colleagues, a codified system such as the one proposed in Chapter 2 results extremely useful to avoid misunderstandings that necessarily mirror in longer and cumbersome procedures. This becomes a must when dealing with multicomponent data.

- HVSR acquisitions: the length of the recording should be consistent with the goal of the survey. When dealing with seismic-hazard studies, acquisitions should be 30 min at least and, when possible, repeated under different meteorological conditions. When possible, avoid working on stiff covers (asphalt, etc.).

- While inverting the data, consider the approach that better suits the data (their complexities) and the goals. Depending on that, the skilled person can decide to consider the modal dispersion curves (in general terms, this is not recommended), the apparent (*effective*) dispersion curve, or the FVS approach. Typically, *effective* dispersion-curve analysis better suits passive data, while the velocity spectra obtained from active acquisitions are often better handled through the FVS approach. In any case, some preliminary direct modeling (see Chapters 5 and 6) is always extremely useful to properly fix the best inversion parameters.

Eventually: do not ever forget or undervalue that the success of a survey is first of all proportional to the *know how* of the people involved and not to the cost of the equipment or other irrelevant acquisition parameters. After all, in the end, we should always be simply humble and reasonable enough to do something only when we (perfectly) know what we are doing.

A test of our understanding is whether we can apply it in practice

Robert Fripp

Appendix—A Collection of Commented Case Studies

The difference between theory and practice is always bigger in theory than in practice.

Folklore

In the following, a series of case studies is presented with the aim of providing a practical evidence of both opportunities and problems to face while analyzing surface-wave propagation and dispersion.

Since the purpose of the presented case studies is to warn and help the practitioner while analyzing complex datasets, in order to properly understand the presented case studies, it is necessary to have gone through all the aspects illustrated in the previous chapters.

This is the reason why, with few exceptions, simple and trivial data that could be treated while adopting simple assumptions (such as, for instance, the idea of dominating fundamental mode) were avoided.

As a matter of facts, actual field datasets can be rarely treated following simple paradigms and, to result in sound analyses, the understanding of the velocity spectra (the basic "object" to consider while analyzing surface waves) has often necessarily to pass through approaches different than the simple modal dispersion curve identification and analysis.

It must be underlined that most of data presented in the following pages were acquired by third parties under nonoptimal conditions and, in some cases, with inexpensive equipment.

In fact, actual field conditions and datasets are generally favorable only in case of commercial fliers and academic papers for which the authors can carefully choose the data capable of supporting their *theses* (that sometimes apply only under specific conditions).

On the opposite side, most of the times, in the real world where professionals work and for which they are required to provide reliable subsurface models, data appear from messy to inscrutable.

The following pages intend to be representative of the real world and serve as a link between all the theoretical aspects described in the previous chapters and the field practice.

CASE STUDY 1

A Simple ZVF Analysis for Geotechnical Purposes

Focus: a first example of Rayleigh-wave data processing while considering both classical modal dispersion curves and *Full Velocity Spectrum (FVS)* inversion. Goal: identification of the depth of the *bedrock* in the framework of a geotechnical study aimed at understanding a differential subsidence that caused some cracks on a building.

In this case, we will consider the data recorded by considering a simple and so-to-speak classical acquisition: an 8 kg-sledgehammer was in fact used to generate the impulse then recorded be means of 23 vertical geophones. In other words, by adopting the nomenclature presented and suggested in Chapter 2, in this case we will only deal with the classical ZVF component.

Figure A1.1 reports the acquired field traces. From the practical point of view it is important to point out that, considered the length of the array (maximum offset 72 m), the acquisition time (1 s) resulted barely sufficient to fully catch and image the Rayleigh-wave dispersion: at the far offsets late arrivals are recorded but a slightly longer

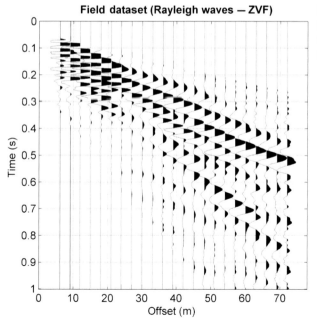

Figure A1.1 Field dataset (seismic traces) obtained while adopting vertical geophones and vertical source (classical ZVF component).

Surface Wave Analysis for Near Surface Applications
ISBN 978-0-12-800770-9, http://dx.doi.org/10.1016/B978-0-12-800770-9.15001-0

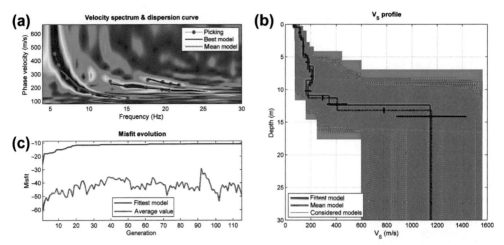

Figure A1.2 Result of the inversion performed while considering the classical approach (picking and inversion of the modal dispersion curves): (a) reported the observed phase velocity spectrum (colored background) and the picked and inverted modal curves; (b) on the right the obtained V_S profiles for the mean and the "best" (the one having the smallest misfit) models (green area represents the adopted search space); (c) the evolution over the misfit values (inversion was accomplished via evolutionary algorithms).

acquisition time would have been preferable even if the velocity spectrum reported in Figure A1.2 appears anyway absolutely well characterized. In this figure are reported the results of the classical approach consisting in the picking of the modal dispersion curves and their successive inversion (in this case we considered and picked the first three modes—quite evident in the observed spectrum). Please consider that, as widely underlined in the pages devoted to the inversion procedures (Chapter 5), this way we are not going to invert the *data* but rather our *data interpretation* (i.e., our picking). It means that in case our picking (i.e., our interpretation/understanding of the velocity spectrum) is wrong, our inversion process would result meaningless even if we obtain a very small misfit.

In this specific case the considered dataset does not show any serious interpretative issue but—in general terms—that depends on the data and on the experience of the practitioner.

The best model (the one with the smallest misfit) and the mean one (in this case defined according to a *Marginal Posterior Probability Density* computation—see Dal Moro et al., 2007) result absolutely similar and this is quite normal (in case a relevant discrepancy between the two models would appear, that would suggest that the inversion procedure was somehow inappropriate).

A further inversion process was performed according to the FVS approach (see Chapter 6) with the results reported in Figure A1.3. In this case there was no

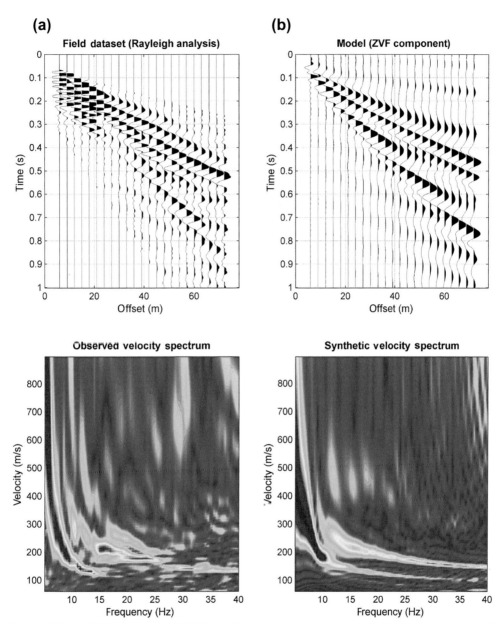

Figure A1.3 *Full Velocity Spectrum* (FVS) inversion: left column (a) reports the field seismic traces and the phase velocity spectrum of the ZVF component; right column; (b) the same data for the identified model (see Figure A1.4). ZVF: Vertical component of Rayleigh waves obtained considering a vertical-impact source (see Chapter 2).

dispersion-curve picking and was inverted the entire phase-velocity spectrum, without any prior interpretation in terms of dispersion curves.

Please notice that, in the velocity spectrum of the identified model (right column in Figure A1.3), the distribution of energy among the different modes is reasonably well represented.

In this case—being the dataset quite simple—there is no relevant difference between the results obtained following the classical and the FVS approaches, even if the FVS inversion suggests a slightly deeper bedrock (see Figure A1.4). Only a joint approach with further data would have been capable of better constraining the solution and providing a precise depth of it (see pages devoted to the joint inversion in Chapter 5).

Eventually, Figure A1.5 reports a sketch of the site which indicates the position (S3) of the borehole that confirmed the correct identification of the bedrock depth.

Figure A1.4 V_S model identified via Full Velocity Spectrum inversion.

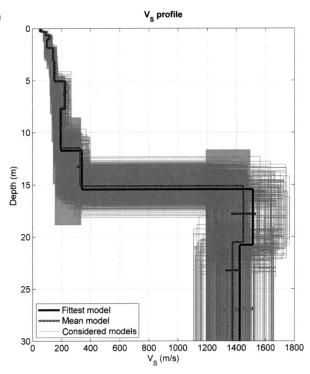

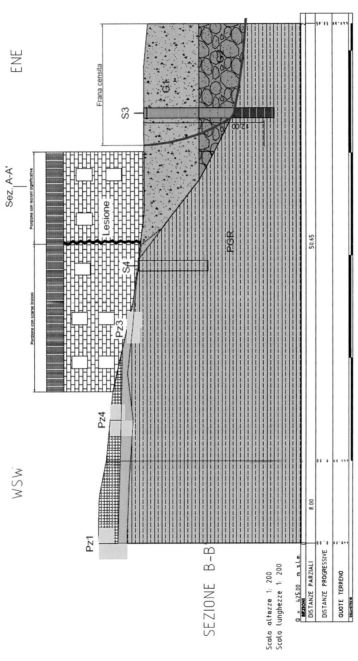

Figure A1.5 Sketch of the geological model identified to explain the differential subsidence that caused some cracks on the building.

CASE STUDY 2

A Simple (but Educational) Case Study

Focus: an easy but comprehensive piece about the joint analysis of Rayleigh and Love waves, also jointly with P-wave refraction travel times, H/V spectral ratio, and the holistic analysis of Rayleigh waves according to group velocities and *radial-to-vertical spectral ratio* (RVSR) (from a purely active acquisition). A reclamation area in northeast Italy (Figure A2.1).

Figure A2.1 The site (a small basin): a soft-sediment sequence lies over a calcarenite formation. How deep will we find the *bedrock*?

Surface Wave Analysis for Near Surface Applications
ISBN 978-0-12-800770-9, http://dx.doi.org/10.1016/B978-0-12-800770-9.15002-2

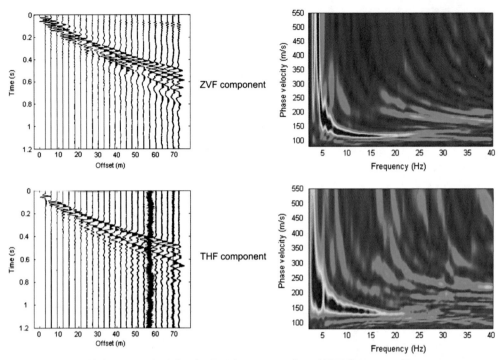

Figure A2.2 Field data (raw data) for the ZVF (upper panel) and THF (lower panel) components.

Figure A2.2 reports the raw data and the phase-velocity spectra of the ZVF and THF components (i.e., the vertical component of Rayleigh waves and Love waves).

Some high-frequency noise is clearly present especially at the distant offsets (being interested in evaluating also the attenuation, the gain/amplification was kept at the same value for all the traces).

Removing this (incoherent) noise is possible by simply applying a low-pass filter but of course such an operation will not "improve" the velocity spectra.

In Figure A2.3 are shown the data after the application of such a low-pass filter. The red polygon in the upper panel puts in evidence a P-wave refraction event (apparent velocity of about 1500 m/s) that we can remove from the data.

To point out the (small) effect of the removed refraction event on the phase-velocity spectrum, in Figure A2.4 are shown the cleaned seismic traces (and the corresponding phase velocity spectrum) after having kept only the signals related to the Rayleigh waves (Figure A2.5).

Figure A2.6 presents the RVF traces (i.e., the radial component). Compared to the ZVF component (Figure A2.3), P-wave refraction is seemingly absent. Actually if we apply some *Automatic Gain Control*, we can eventually highlight the refraction event (Figure A2.7) having the same slope (i.e., velocity) as the one apparent on the ZVF component reported in Figure A2.3 (upper panel).

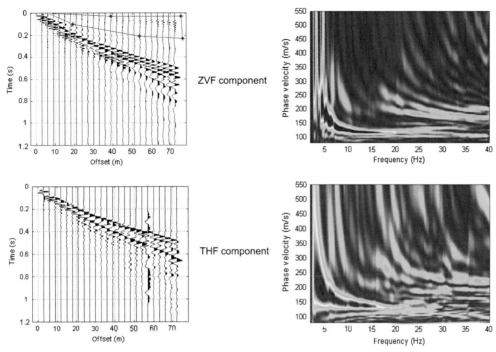

Figure A2.3 Field data after a low-pass filter applied to remove the high-frequency noise. The red polygon highlights a P-wave refraction (slope equals to about 1500 m/s).

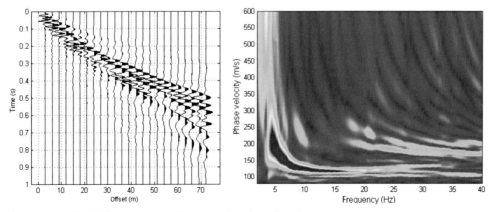

Figure A2.4 The ZVF component after removing the refraction event.

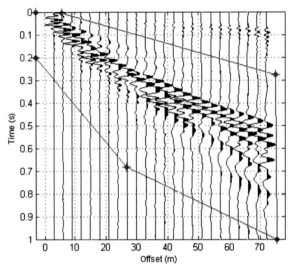

Figure A2.5 Removing the P-wave refraction event from the ZVF component (only the data within the polygon are kept).

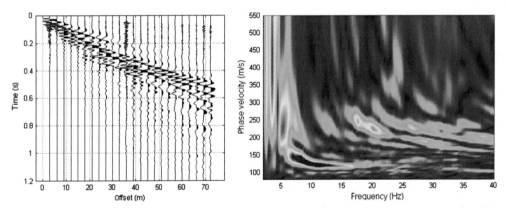

Figure A2.6 RVF component: please notice that the P-wave refraction (quite apparent in the ZVF component) seems to be almost absent and that the energy distribution among the different modes is slightly different from the ZVF component (compare with Figure A2.4).

It might be interesting to see why the large-amplitude refraction evident on the ZVF component seems to have a much smaller amplitude on the RVF component. The basic reason is actually quite simple and just requires recalling the well-known Snell's equation describing the refraction process (because of its popularity, the meaning of the symbols does not need any further explanation):

$$\frac{V_1}{V_2} = \frac{\sin(\theta_1)}{\sin(\theta_2)}$$

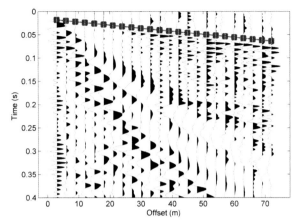

Figure A2.7 RVF component after the application of an *Automatic Gain Control* capable of putting in evidence the same refraction which, on the ZVF component (Figures A2.3 and A2.5), is much clearer (larger amplitude).

If we now consider two layers (characterized by a large difference in the velocity values—e.g., 250 and 1500 m/s) the consequence is straightforward: the critical angle (i.e., the angle formed by the emerging refracted wave and the vertical) is going to be quite small (less than 10°). That clearly means that the radial component of this vector (i.e., the amplitude of the radial component of the refracted wave) is much smaller than the vertical one (i.e., the amplitude of the vertical component of the refracted wave).

We can now put together all these facts and summarize the point: V_S values of the shallow layers (easily retrieved from surface waves) are quite low (around 100 m/s), while P-wave refraction indicates high P-wave velocities (≈ 1500 m/s) already at a depth of about 2 m and, furthermore, its amplitude is much larger on the vertical component than on the radial component.

All these features are consistent with a very clear situation: the presence of the water table (whose V_P is notoriously around 1500 m/s) in a soft-sediment sequence (whose V_S values are typically around 100–200 m/s).

It can be recalled that, for this kind of porous materials, compressional wave velocities are strongly influenced by the water (while a shallow dry sand has usually V_P values around 300–400 m/s, the presence of water will increase V_P up to 1500–1800 m/s), while shear-wave velocities undergo only minor changes (water tends to slightly lower them - see Paragraph 7.2.2)—this is why there is no refraction on the THF component.

Because of the very soft (thus slow) sedimentary cover, the *multichannel analysis of surface waves* data were unable to strictly constrain the deep layers (namely the contact between the sediments and the bedrock). In order to do it, we thus also considered the H/V spectral ratio.

Figures A2.8 reports the observed horizontal-to-vertical spectral ratio (HVSR) computed while considering two different spectral smoothing: while the general features are inevitably preserved, it is nevertheless interesting to point out the fact that while using a smaller amount of smoothing, the peak cantered around 1.8 Hz divides into two sub-peaks and that this evidence is somehow mirrored in the RVSR curve (the *RVSR* obtained via active acquisition) considered later on in the joint analysis performed while also considering the group-velocity spectra.

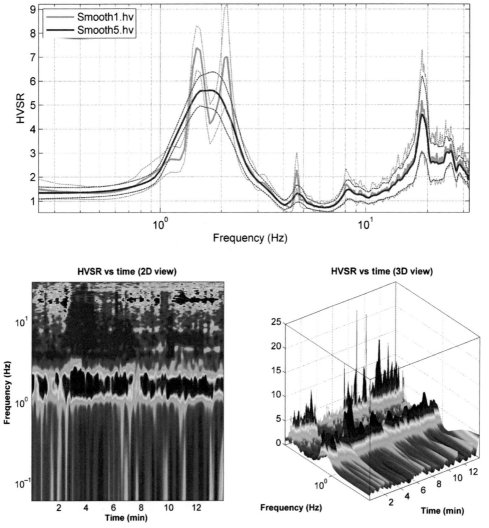

Figure A2.8 Observed horizontal-to-vertical spectral ratio (HVSR): in the upper panel the average HVSR curves (computed considering 1% and 5% smoothing) and, in the lower panel, its temporal continuity over the considered recording (about 14 min).

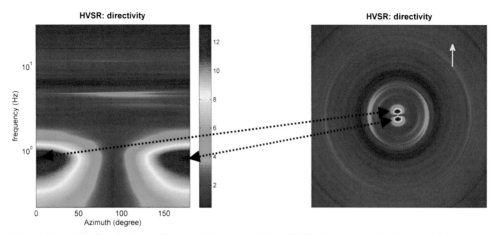

Figure A2.9 A further horizontal-to-vertical spectral ratio (HVSR) dataset acquired over a slope of the surrounding hills while the *north–south* axis of the 3-component geophone was aligned along the maximum slope so that the slope is represented by the *zero* azimuth: the topographic effect is apparent. The colors refer to the HVSR values. The center of the circular plot on the right represents the minimum considered frequency (0.2 Hz) that increases moving outward.

In order to show the possible effect of the topography on the observed HVSR, a further dataset was acquired over the slope of the surrounding hills (Figure A2.1). The directivity analysis reported in Figure A2.9 shows quite clearly the effect of the topography (the north-south axis of the geophone was aligned with the slope).

Usually, observed HVSR is essentially determined by the surface waves dominating the background microtremor field (see Chapter 4). Nevertheless, in some cases, the fundamental resonance frequency is better or equally described by body waves (e.g., Albarello and Lunedei, 2010) and this seems one of those cases.

Figure A2.10 reports the solution of the site obtained while considering both Rayleigh and Love waves, jointly with HVSR (for further details see Dal Moro, 2010).

The solution of this site (a simple small basin where a soft-sediment sequence lies over the solid bedrock) was also pursued by holistically analyzing the group velocities of the vertical and radial components of Rayleigh waves together with the RVSR (see Section 2.2.2). The simple acquisition setting (a classical vertical-impact sledgehammer source and a single 3-component geophone 50 m away from the source) provided the three objective functions then used to infer the subsurface model. From the practical point of view, the reader should consider that if the RVSR is modeled while considering Rayleigh waves only, it is important to remove from the field traces the contribution of refraction events (or any other signal not related to the surface waves) and/or to fix a sufficiently-large *offset* (in general terms, because of the attenuation due

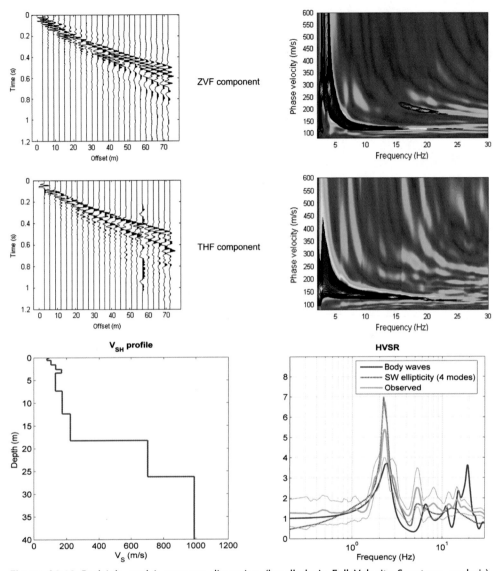

Figure A2.10 Rayleigh- and Love-wave dispersion (handled via Full Velocity Spectrum analysis), horizontal-to-vertical spectral ratio (HVSR) (modeled according to both surface and body waves), and identified V$_S$ model. See also Dal Moro (2010).

to the geometrical spreading, body waves decrease their amplitudes more than surface waves).

The outcomes of the multi-objective inversion are reported in Figures A2.11 and A2.12 where the overall consistency of the inversion procedure (see Chapters 5 and 6) and good match between field and synthetic data are apparent.

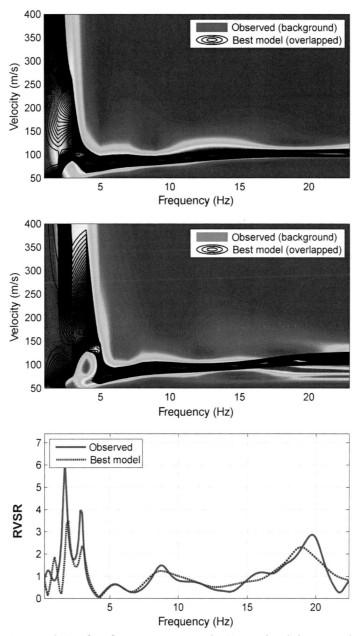

Figure A2.11 Joint analysis of surface-wave group velocities and radial-to-vertical spectral ratio (RVSR) [a patent-pending methodology]: background colours represent the field-dataset group-velocity spectra of the vertical (a) and radial (b) components of Rayleigh waves, while the overlaying contour lines the synthetic group velocity spectra of the identified model; (c) the red and blue curves report the synthetic and observed RVSRs, respectively. See also Figure A2.12.

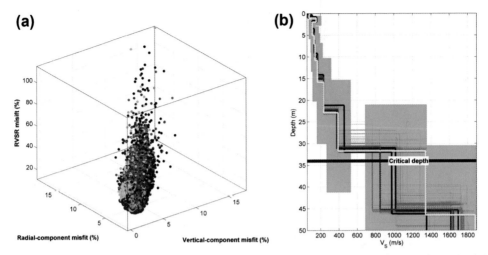

Figure A2.12 Joint analysis of surface-wave group velocities and radial-to-vertical spectral ratio (RVSR): (a) misfit values for the three objective functions; (b) retrieved V_S models. See also Figure A2.11. The *critical depth* indicates two thirds of the considered offset (50 m) [see paragraph 2.2.2].

CASE STUDY 3

Inverse Dispersion by the Book

Focus: a rare example of extremely apparent inverse dispersion of Rayleigh waves solved via Full Velocity Spectrum (FVS) analysis.

The presented dataset represents a one-of-a-kind case of apparent inverse dispersion, which can actually result quite misleading. In fact, in spite of the continuity of the signal in the velocity spectrum, the apparent dispersion curve is actually the result of several higher modes whose energy is so large because of a massive stiff layer that creates a noteworthy velocity inversion.

Unfortunately, since the dataset was not acquired following the author's recommendations regarding the multicomponent analysis, only the classical ZVF component (vertical geophones and vertical-impact source) is available.

Figure A3.1 presents the acquired traces and the computed phase-velocity spectrum that shows a very distinctive inverse apparent dispersion.

An FVS inversion was performed with the results summarized in Figure A3.2. Actually, since the apparent dispersion curve is unique (please compare with the velocity spectrum presented in Figure A6.4, where three apparent and distinctive modes did not allow to adopt an approach based on the *effective* dispersion curve), in this case, an inversion procedure based on a picked *apparent* (or *effective*) dispersion curve (Tokimatsu et al., 1992) would have been equally viable.

Together with the surface-wave analysis, Figure A3.2 also reports the V_P model (V_P were adjusted/modeled starting from the values obtained from the FVS inversion) and the P-wave refraction arrival travel times.

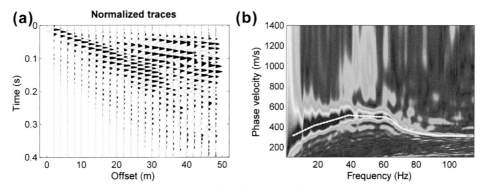

Figure A3.1 Field dataset: (a) seismic traces (ZVF component), (b) phase-velocity spectrum (dotted line indicates the *apparent*, or *effective*, dispersion curve).

Surface Wave Analysis for Near Surface Applications
ISBN 978-0-12-800770-9, http://dx.doi.org/10.1016/B978-0-12-800770-9.15003-4

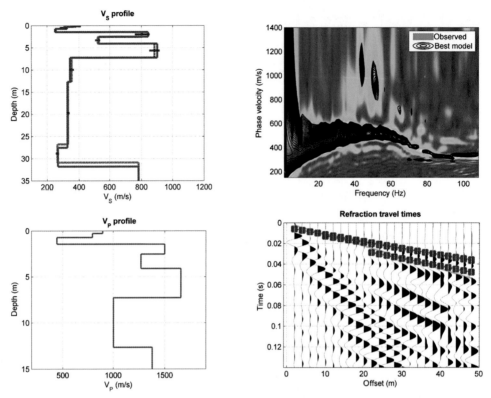

Figure A3.2 Upper panel: V_S model and field (background colors) and synthetic (black contour lines) phase-velocity spectra; lower panel: V_P model and P-wave refraction travel times computed by considering a ray-tracing procedure capable of taking into account low-velocity channels (red squares relate to refraction events while the green line represents the direct wave) (shown only the uppermost layers).

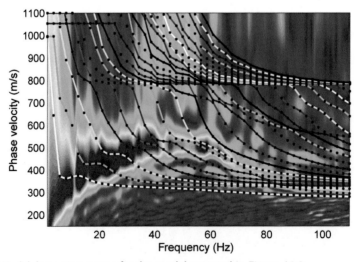

Figure A3.3 *Modal* dispersion curves for the model reported in Figure A3.2.

Two crucial remarks to point out:

1. Once again (see also for instance Figures 3.2 and 3.3) the continuity of a signal in the velocity spectrum does not correspond to a unique mode (multiple modes can merge in a continuous signal without the evidence of any "mode jump"—see Chapter 3) (the modal dispersion curves for the identified model are presented in Figure A3.3).

2. In order to properly face this sort of datasets, only an approach based on the *effective* dispersion curve or on the FVS analysis (which is somehow an extension of it) is feasible. An approach based on the modal dispersion curves would not be possible simply because of the impossibility of differentiating the single modes.

Useless to underline that (as always—see Chapter 5) the acquisition of additional components (e.g., RVF, THF, or the horizontal-to-vertical spectral ratio) would have been extremely useful to further constrain the solution and further validate the presented analyses.

CASE STUDY 4

When the Joint Analysis of Love and Rayleigh Waves Is Necessary

Focus: joint inversion of Rayleigh- and Love-wave modal dispersion curves via *multiobjective evolutionary algorithms* (MOEA) and evaluation of the Pareto-front symmetry (see Chapter 5) while also considering horizontal-to-vertical spectral ratio (HVSR) and *cone penetration test* (CPT) data.

The presented *dataset* was acquired for geotechnical purposes in Tuscany (Italy) while using horizontal geophones only so that, together with Love waves, Rayleigh waves were acquired according to their radial component. Because of logistic limitations, the maximum possible offset is 36.5 m (see Figure A4.1). It can be easily noted that in the 20–40 Hz frequency range (pertaining to the shallowest layers), phase velocities of the signal dominating the Rayleigh-wave velocity spectrum are clearly higher (about 400 m/s) with respect to the ones shown by the Love-wave velocity spectrum (about 250 m/s).

Although the theoretical dispersion curves of Rayleigh and Love waves are clearly different, such a large difference should immediately sound as a warning because it clearly suggests that the two signals pertain to different modes.

With the aim of illustrating the use of Pareto-front analysis in a joint inversion procedure based on the classical modal dispersion curve picking, we analyzed the data according to two different interpretative hypotheses.

Hypothesis 1 (the one that would have been likely followed by nonexperts): the signal dominating the Rayleigh-wave velocity spectrum in the 20–50 Hz is interpreted as belonging to the fundamental mode.

Hypothesis 2: the same signal is attributed to the first higher mode while the small signal in the 25–30 Hz range with a phase velocity of about 250 m/s to the fundamental mode (please note that we could also avoid this second "attribution" thus considering the first higher mode only).

In both cases, Love-wave dispersion is associated to the fundamental mode (please consider all the evidences illustrated in Chapter 3 and all the reported case studies).

Results of the joint inversion performed while considering the *first hypothesis* are reported in Figures A4.2 and A4.3. The two most important facts to underline:

1. The dispersion curves of the Pareto models (i.e., the "best" models) do not match the picked ones (Figure A4.2) since Rayleigh-wave dispersion curves tend to stay below (i.e., to be slower than) the picked one while for Love waves the situation is contrary

Surface Wave Analysis for Near Surface Applications
ISBN 978-0-12-800770-9, http://dx.doi.org/10.1016/B978-0-12-800770-9.15004-6

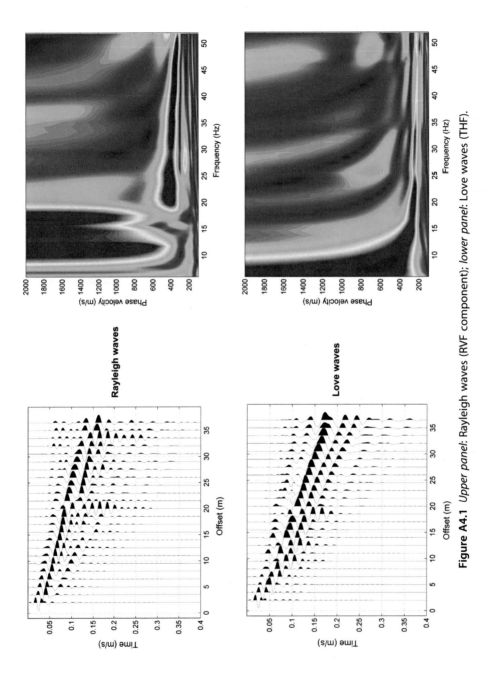

Figure A4.1 *Upper panel:* Rayleigh waves (RVF component); *lower panel:* Love waves (THF).

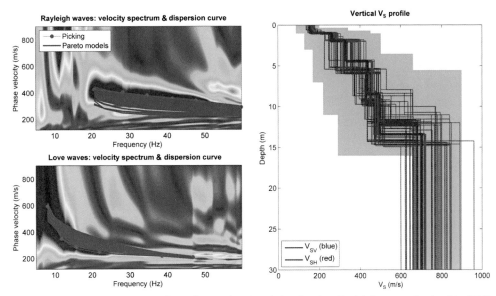

Figure A4.2 First interpretative hypothesis: the signal dominating Rayleigh waves in the 20–55 Hz frequency range (phase velocity approximately equal to 400 m/s) is attributed to the fundamental mode.

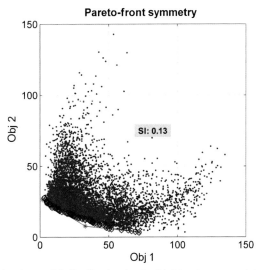

Figure A4.3 First hypothesis: model distribution in the biobjective space (obj 1 represents Rayleigh-wave misfit, obj 2 relates to Love waves). The cloud of models does not point toward the [0, 0] *utopia point* and the Pareto-optimal models (red circles) are not symmetric with respect to the *universe* of evaluated models (symmetry index (SI) equal to 0.13—see Dal Moro and Ferigo, 2011).

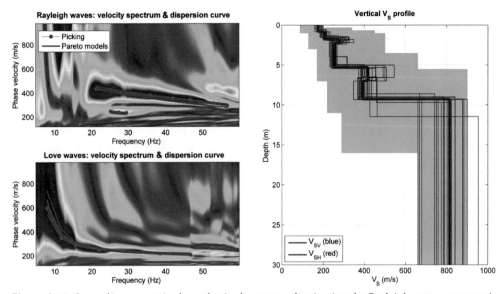

Figure A4.4 Second interpretative hypothesis: the energy dominating the Rayleigh-wave spectrum in the 20–55 Hz frequency range is interpreted as belonging to the first higher mode.

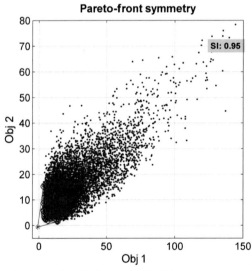

Figure A4.5 Second hypothesis: model distribution in the biobjective space. Now the cloud of models points toward the *utopia point* and the Pareto-optimal models (red circles) are highly symmetric with respect to the *universe* of evaluated models (symmetry index (SI) equal to 0.95).

(dispersion curves of the Pareto-front models tend to be higher—faster—than the picked one).

2. The model distribution in the biobjective space (Figure A4.3) is clearly highly unbalanced: the cloud of models does not point toward the [0, 0] *utopia point* and the Pareto-front models are not symmetric when compared to the *universe* of considered models (for details, please see Chapter 5 and the mentioned literature).

Figures A4.4 and A4.5 report the main outcomes of the joint inversion performed while considering the *second interpretative hypothesis* (the energy dominating the Rayleigh-wave velocity spectrum is now interpreted as belonging to the first higher mode).

Now the dispersion curves of the models belonging to the Pareto front *perfectly* overlap with the picked dispersion curves (Figure A4.4) for both Rayleigh and Love waves. Moreover, the model distribution in the biobjective space consistently points toward the *utopia point* (Figure A4.5).

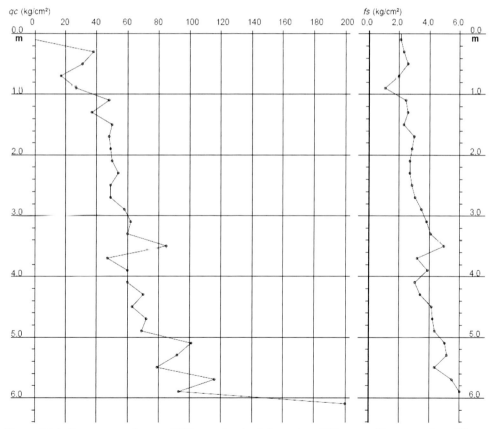

Figure A4.6 *Cone penetration test* (CPT): at a depth of about 6 m, CPT stopped because of a stiff layer. Compare with the horizon at a depth of about 5.5 m put in evidence by the joint analysis presented in Figure A4.4.

Furthermore, now the V_S profiles of the Pareto-front models (right panel in Figure A4.4) appear extremely so-to-say *focused* (compare with the V_S profiles reported in Figure A4.2) showing two main horizons: the first at a depth of about 5.5 m and the second at about 10 m.

All these facts indicate an overall mathematical consistency of this second interpretative hypothesis (for details about the paradigms to adopt while evaluating MOEA analyses, please refer to Dal Moro and Ferigo (2011)), which is supported by two further facts:

CPT DATA

Figure A4.6 reports the CPT (Cone Penetration Test) that stopped at about 6 m because of a stiff layer that, in the light of the surface-wave analysis, can be interpreted as a gravel-like material, while the real *bedrock* is actually deeper (10 m depth—see the V_S profile reported in Figure A4.4).

HVSR DATA

A further validation of the overall consistency of the above-presented analyses is given by the *HVSR* acquired on the site. Figure A4.7 reports the observed and modeled HVSR curves (the synthetic curve refers to the average model obtained from the final Pareto-front models reported in Figure A4.4).

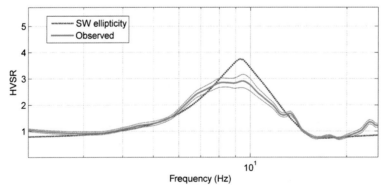

Figure A4.7 Observed and modeled (average model from the Pareto-front models obtained from the second interpretative hypothesis) *horizontal-to-vertical spectral ratio* (HVSR). Modeling performed according to Arai and Tokimatsu (2004), while considering the combined effect of Rayleigh and Love waves (Surface Waves, SW) on the observed HVSR.

CASE STUDY 5

Joint Analysis of Rayleigh-Wave Dispersion and P-Wave Refraction

Focus: an example of Rayleigh-wave *mode splitting* due to the presence of a superficial V_S inversion and high Poisson ratios related to saturated sands. A joint analysis of Rayleigh-wave dispersion and P-wave refraction travel times.

The dataset was acquired nearby a sandy beach in the most classical way, that is vertical-component geophones and vertical-impact sledgehammer or, following the simple and efficient nomenclature suggested in Chapter 2: ZVF.

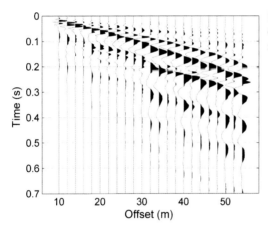

Figure A5.1 Field dataset: seismic traces (ZVF component) and computed phase-velocity spectrum.

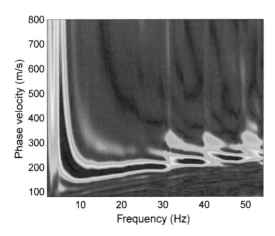

Surface Wave Analysis for Near Surface Applications
ISBN 978-0-12-800770-9, http://dx.doi.org/10.1016/B978-0-12-800770-9.15005-8

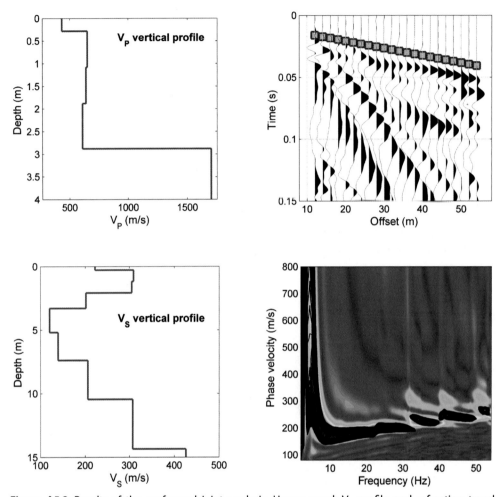

Figure A5.2 Results of the performed joint analysis. Upper panel: V_P profile and refraction travel times; lower panel: V_S model and observed and computed phase-velocity spectra (background color shows the field data, overlaying contour lines show the synthetic velocity spectrum).

Data (and computed phase-velocity spectrum) are presented in Figure A5.1 and show that very typical *mode splitting* presented in Chapter 3 (please note the similarities with respect to both seismic traces and velocity spectrum presented in Figure 3.4 and 3.5) where it was highlighted the fact that this kind of phenomenon is typically the evidence of two facts: a superficial relatively-stiff layer and a deeper softer stratum characterized by very high Poisson moduli.

Results of the joint analysis of P-wave refraction travel times and ZVF dispersion (Dal Moro, 2008) were performed with the results summarized in Figure A5.2 (dispersion was considered following the Full Velocity Spectrum approach illustrated in Chapter 6 and

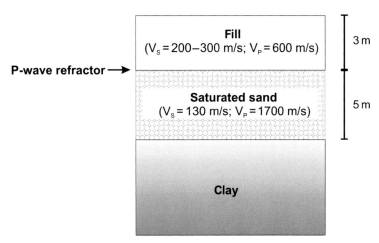

Figure A5.3 Stratigraphic model identified by considering the results of the joint analysis presented in Figure A5.2.

used in a number of presented case studies). The stratigraphic model that was identified on the basis of such results is presented in Figure A5.3: after a few meters (about 3 m) of fill material, is present an autochthonous saturated sandy layer characterized by lower shear-wave velocities (see also Paragraph 7.2.2) but higher V_P values (due to the water saturation). It is just this peculiar sequence that creates the *mode splitting* quite apparent in the velocity spectrum reported in Figure A5.1 and the P-wave refraction. Eventually, after few meters of sand, clay sediments (sampled by local borehole) slowly determine the observed V_S increase with depth.

CASE STUDY 6

A Comprehensive Survey in the Swiss Alps

G. Dal Moro[1], L. Keller[2]

[1] *Eliosoft*, Udine, Italy; [2] roXplore Gmbh (Amlikon-Bissegg, Switzerland)

Focus: an *all-inclusive* example of joint analysis of Rayleigh- and Love-wave dispersion performed according to several active and passive techniques and both according to phase and group velocities. Results are also compared with vertical seismic profile (VSP) log.

The comprehensive dataset here considered *dataset* was acquired for the characterization of an area where a borehole short-period seismic station and a surface strong-motion sensor are installed and run by the *Swiss Seismological Service* with the final goal of assessing the local seismic response.

The area (a Molassic hill relief—see Figure A6.1) consists mainly of siltstone and marl with some minor layers of fine-grained sandstone.

Two field datasets were acquired along two nearly perpendicular 95 m-long lines, about 30 m away from a borehole where VSP (Vertical Seismic Profile) data were also collected (see also Dal Moro and Keller, 2013).

The use of both vertical and horizontal geophones allowed acquiring a complete set of multichannel data consisting of (see also nomenclature described in Chapter 2):

- vertical component of Rayleigh waves (vertical sledgehammer and vertical geophones—ZVF)

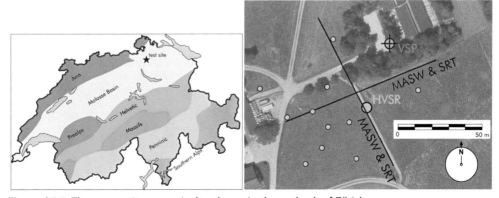

Figure A6.1 The survey site: an agricultural area in the outback of Zürich.

Surface Wave Analysis for Near Surface Applications
ISBN 978-0-12-800770-9, http://dx.doi.org/10.1016/B978-0-12-800770-9.15006-X

Figure A6.2 Field data acquisitions: shear and Love waves efficiently (and economically) produced by means of a sledgehammer and a horizontal wooden beam. To improve the coupling, a couple of spikes nail the beam to the soil.

- radial component of Rayleigh waves (vertical sledgehammer and horizontal/radial geophones—RVF)
- Love waves (wooden beam (Figure A6.2) and transversal geophones—THF)

Furthermore, a set of data acquired by means of a calibrated 3-component (3C) geophone were acquired both according to passive (horizontal-to-vertical spectral ratio) and active procedures.

Eventually, ETH (the *Eidgenössische Technische Hochschule* of Zürich) members (Valerio Poggi and Donat Fäh) also acquired passive data by means of circular arrays consisting of a series of low-frequency (5s eigenperiod) 3C sensors whose data were then analyzed according to Poggi and Fäh (2010) to image surface-wave dispersive properties at low frequencies.

Let us start with some comments about the active (MASW) data. Please notice (Figure A6.3) the differences in the ZVF and RVF phase-velocity spectra: fundamental mode is evident in the RVF component (see the signal in the 10—20 Hz frequency range, characterized by a 400 m/s phase velocity) while the effective dispersion curve of the ZVF component is by far more complex and shows no evidence of the fundamental mode.

Final outcomes of the performed joint analysis are summarized in Figures A6.4 and A6.5: the first reports the radial component of Rayleigh waves together with the H/V spectral ratio and the obtained V_S profile, while in the second are presented the data pertaining to Love waves.

Love-wave dispersion curve obtained by Poggi and Fäh via *frequency—wavenumber* analysis (Poggi and Fäh, 2010) appears remarkably consistent with the velocity spectrum obtained by means of active (*multichannel analysis of surface waves*) acquisitions (and lacking of low frequencies because of the used 10 Hz horizontal geophones). Data reported in

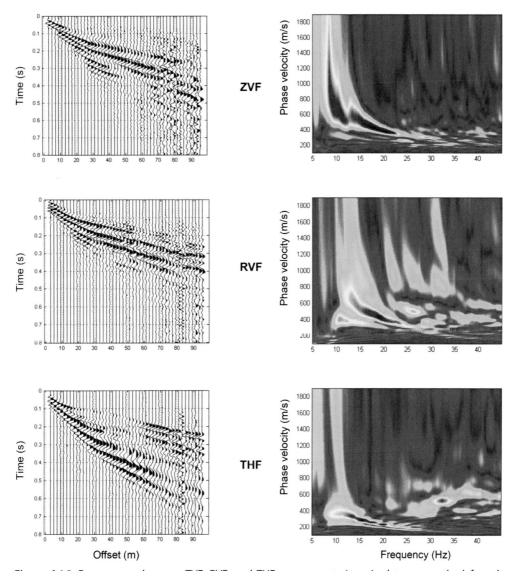

Figure A6.3 From top to bottom: ZVF, RVF, and THF components (acquired traces on the left and corresponding phase-velocity spectra on the right). Velocity spectra of the RVF and THF components for frequencies lower than 10 Hz are poorly defined because the adopted horizontal geophones were characterized by an *eigen* frequency of 10 Hz (common geophones typically used for SH-wave (the Horizontal component of the Shear waves) refraction surveys).

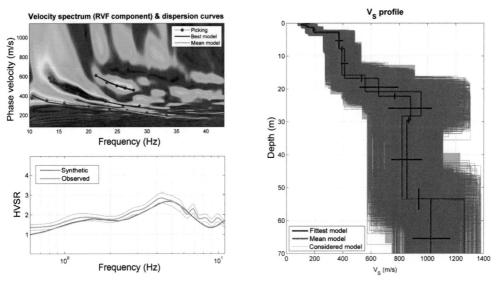

Figure A6.4 Joint analysis of Rayleigh-wave (radial component—RVF) dispersion and horizontal-to-vertical spectral ratio (HVSR). The phase-velocity spectrum of the RVF component shows three apparent modes (including the fundamental one).

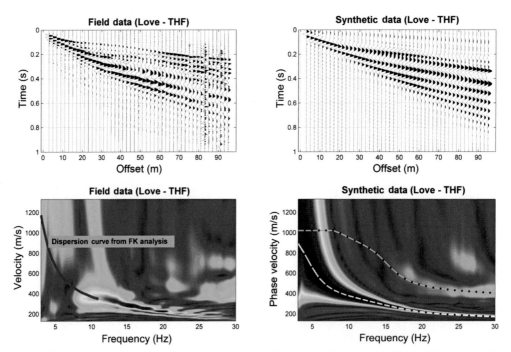

Figure A6.5 Love waves: on the upper panel the field and synthetic (same V_S model reported in Figure A6.4) traces; on the lower panel the velocity spectrum from the active acquisition with, over-laying, the dispersion curve obtained via *frequency—wavenumber* analysis (Poggi and Fäh, 2010). Synthetic traces are computed via modal summation. The overall consistency with the observed data is apparent.

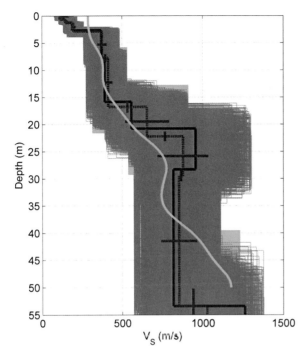

Figure A6.6 V_S profiles obtained from the VSP (Vertical Seismic Profile) analysis (green) and the joint analysis of surface-wave dispersion and HVSR (blue and red lines represent the best and the mean models, respectively – see also Figure A6.4).

Figure A6.6 allow to compare the V_S profile obtained through the VSP analysis and the one resulting from the joint analysis of surface-wave dispersion and HVSR (the overall agreement is apparent).

A further active acquisition was performed while considering a single 3C (calibrated) geophone at a distance of 70 m from the source. Collected data were analyzed by considering the group-velocity spectra of both the vertical and radial components and to the *radial-to-vertical spectral ratio* (RVSR) (see Paragraph 2.2.2 and case studies 2 and 8).

The recorded vertical and radial components were used to compute the respective group-velocity spectra and then inverted (following the Full Velocity Spectrum approach—see Chapter 6) jointly with the RVSR according to a multiobjective procedure (in this case the number of objective functions is clearly three).

Main results are reported in Figures A6.7 and A6.8 where the overall good agreement with the VSP data is quite apparent. Actually, in this sort of joint inversion, it is possible to define several "best" models (Dal Moro, 2014) and, among the things to evaluate in order to understand whether the inversion procedure can be considered meaningful, it is important to verify that they are all congruent (i.e., similar).

Main facts can be summarized in the following points:

1. In this case, the RVF component (i.e., the radial component of Rayleigh waves) computed while considering multichannel data (Figure A6.3), revealed essential to properly understand Rayleigh-wave dispersion in terms of modal dispersion curves: in fact, if considered alone, the *effective* dispersion curve (Tokimatsu et al., 1992) shown

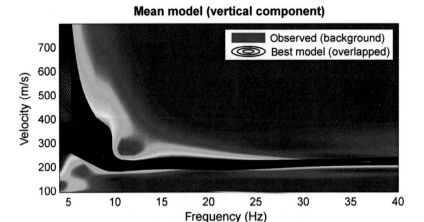

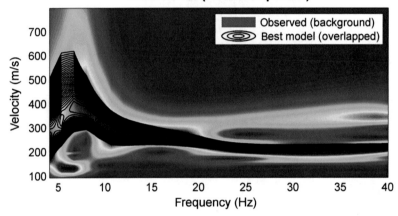

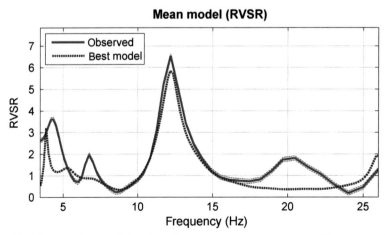

Figure A6.7 Holistic analysis of radial and vertical components (group velocities) of Rayleigh waves jointly with *radial-to-vertical spectral ratio* (RVSR) [a patent-pending methodology]. Colors in the background represent the field data while overlaying black contour lines represent the velocity spectra of the retrieved model.

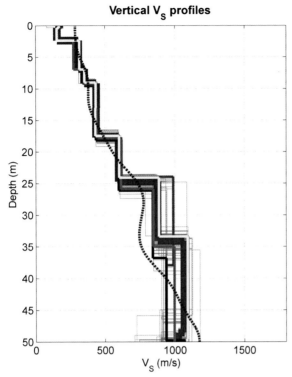

Vertical V$_S$ profiles

Figure A6.8 V$_S$ profiles obtained from vertical seismic profile (black dashed line) and holistic analysis of group-velocity spectra and *radial-to-vertical spectral ratio* (see Figure A6.7): the "best" models result congruent thus providing evidence that the inversion procedure is overall meaningful.

by the ZVF component seriously risks to be misinterpreted, while the RVF velocity spectrum clearly shows the first three modes (including the fundamental one) (compare data presented in Figures A6.3 and A6.4 and Dal Moro and Keller, 2013).

2. The final V$_S$ profile determined by the joint analysis of multichannel passive and active data results in very good agreement with the VSP data.

3. The joint and holistic analysis of the group-velocity spectra of the radial and vertical component performed jointly with the RVSR results quite consistent with the VSP profile down to a depth of about two-thirds of the considered offset (the extremely simple field procedures necessary to perform this sort of analyses must be recalled: a single 3C geophone and a common vertical-impact seismic source).

4. Due to the well-known problems in the identification of shear–wave arrivals in the first few meters, VSP profile down to 3 m significantly overestimates the velocities.

ACKNOWLEDGMENTS

We gratefully acknowledge *Nagra* (*Nationale Genossenschaft für die Lagerung radioaktiver Abfälle*) for the permission of showing the data and Valerio Poggi and Donat Fäh D (ETH, Zurich) who kindly provided us with HVSR and *fk* data.

CASE STUDY 7

Joint Analysis of Rayleigh and Love Waves via Full Velocity Spectrum Analysis

Focus: an example of joint analysis of Rayleigh- and Love-wave dispersion while considering data acquired using only horizontal geophones (RVF and THF components) and processed following the *Full Velocity Spectrum (FVS)* approach.

The *dataset* was acquired for geotechnical purposes at the foothill of a mountain relief (see Figure A7.1), in an area where the superficial sediments are fundamentally represented by silt and clay.

Acquired data are reported in Figure A7.2. Both Rayleigh (radial component acquired using horizontal geophones) and Love waves appear to contain a large amount of higher modes that do not have to be considered as a kind of noise or problem but, on

Figure A7.1 The investigated site.

Surface Wave Analysis for Near Surface Applications
ISBN 978-0-12-800770-9, http://dx.doi.org/10.1016/B978-0-12-800770-9.15007-1

171

Figure A7.2 Field data. *Upper panel*: Rayleigh waves (radial component—RVF); *lower panel*: Love waves (THF).

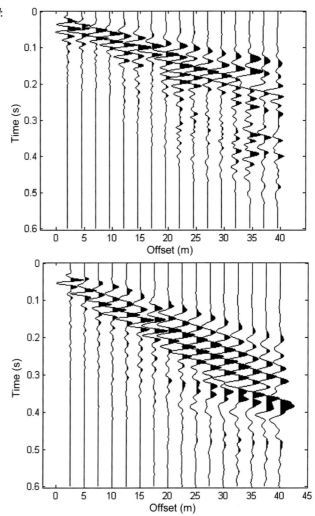

the opposite, as a source of relevant information capable (when properly considered) to better constrain the deepest layers.

Data were processed in order to jointly invert the velocity spectra of the RVF and THF components according to the FVS approach (see Chapter 6) and the Figures A7.3—A7.5 present the results of the performed analysis. The overall consistency of the observed and synthetic data appears clearly remarkable since the correspondence between the observed and inverted velocity spectra is apparent (the retrieved V_S model is reported in Figure A7.6).

It is absolutely noteworthy and illuminating the evidence that the signal dominating the RVF component (radial component of Rayleigh waves) for frequencies higher than

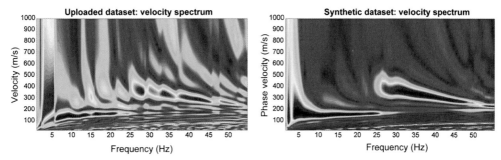

Figure A7.3 Rayleigh waves (radial component—RVF): on the left the field data, on the right the one of the retrieved model (reported in Figure A7.6).

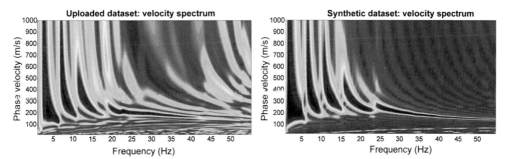

Figure A7.4 Love waves (THF component): on the left the field data, on the right the synthetics of the retrieved model.

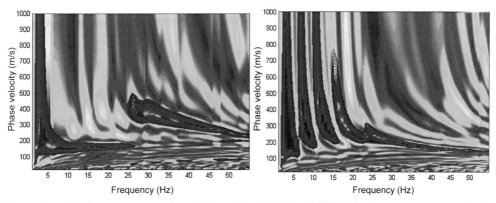

Figure A7.5 Synthetic representation of the results of the joint Full Velocity Spectrum inversion: background colors represent the velocity spectra of the field dataset while overlaying blue contour lines represent the synthetic velocity spectra of the retrieved model reported in Figure A7.6.

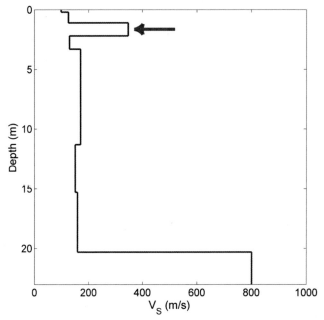

Figure A7.6 Identified V_S model. Red arrow highlights the shallow stiff layer responsible for the observed peculiar mode excitation.

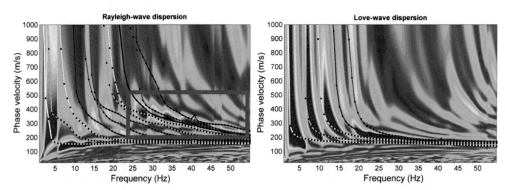

Figure A7.7 On the left the Rayleigh-wave velocity spectrum of the field dataset and, overlaying, the theoretical modal dispersion curves of the identified V_S model. The signal dominating the field data for frequencies higher than 25 Hz (in the pink rectangle) actually relates to a complex combination of higher modes (from the 8th to the 11th) and not (as simplistic assumptions could suggest) to the first higher mode. On the right the same data for the Love waves, showing a simpler behaviour (modes are excited in sequence from the fundamental up to the fourth higher one).

about 25 Hz is not—as some simplistic assumption might induce to imagine—the first higher mode. By plotting (Figure A7.7) the modal dispersion curves of the retrieved model (Figure A7.6), it is in fact clear that the considered signal actually pertains to the combination of a number of higher modes (from the 8th to the 11th), once more

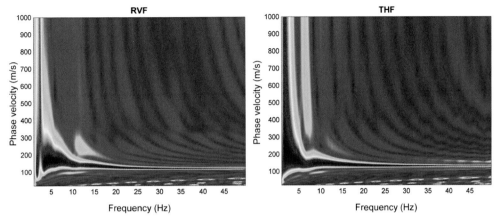

Figure A7.8 Phase velocity spectra for the RVF and THF components of a model identical to the one reported in Figure A7.6, but lacking the stiff layer indicated by the red arrow.

showing that the continuity of a signal does not necessarily mean that a single mode is responsible for it (see all the evidences presented in Chapter 3). It is thus quite clear that this kind of data cannot be solved through the analysis of modal dispersion curves and only an approach based on the effective dispersion curves or on the FVS inversion results actually viable.

It is surely quite remarkable to point out that the stratigraphic feature actually responsible for the peculiar observed distribution of energy among the different modes is a shallow and relatively thin stiff layer (see red arrow in Figure A7.6). In order to highlight this point, we computed the velocity spectra for the same V_S model but now substituting the mentioned stiff layer with a layer having a shear-wave velocity equal to the mean value of the layers over and below it (in other words the stiff layer was simply smoothed out). Resulting velocity spectra for the RVF and THF components are reported in Figure A7.8.

Some final considerations can be summarized as follows:

1. Horizontal geophones can be adopted to efficiently and jointly acquire both Rayleigh (radial component) and Love waves (see Chapter 2).
2. As shown by this example and by the synthetic data reported in Chapter 3, the radial component of Rayleigh waves is not less relevant or applicable than the vertical one. The energy distribution among the different modes can be different with respect to the vertical one (because the respective Green's functions are different) but that does not mean that one component is more or less important than the other.
3. The *Full Velocity Spectrum* approach allows the full exploitation of all the modes present in the field data (and without their interpretation in terms of modal dispersion curves) thus resulting in a very-well constrained inversion procedure capable of providing a robust subsurface V_S model.

CASE STUDY 8

A Civil Engineering Job

Focus: This is an example of geotechnical characterization of the sedimentary cover in a lagoon area. In order to illustrate different possible approaches, data collected at two sites (about 2.5 km distant) are inverted according to different strategies.

In one case (site#2) a joint inversion of *Horizontal-to-Vertical Spectral Ratio* (HVSR) (passive seismics) and Rayleigh-wave group-velocity spectrum (active seismics) is performed and results are eventually compared to simpler single-component *Full Velocity Spectrum* (FVS) inversion of the ZVF group-velocity spectrum only.

For the second considered data set (site#8), active data were holistically inverted by considering the radial and vertical components of Rayleigh waves, jointly with the *Radial-to-Vertical Spectral Ratio* (RVSR).

In the framework of a geotechnical study performed to assess the stability of a river dyke that here and there tends to collapse (Figure A8.1) a series of seismic acquisitions

Figure A8.1 Survey area; the small river along which dyke locally tends to collapse. Shown the two sites considered for the present case study (the river mouth is about 500 m from site#2).

Surface Wave Analysis for Near Surface Applications
ISBN 978-0-12-800770-9, http://dx.doi.org/10.1016/B978-0-12-800770-9.15008-3
177

aimed at defining the local stratigraphy were performed by considering only a single 3-component (calibrated and triggerable) geophone to analyze the Rayleigh wave propagation (active seismics) and the HVSR curve (passive seismics).

In the following, are presented the analyses performed on the data acquired on two sites (site#2 and #8) about 2.5 km one from the other.

SITE#2

Rayleigh-wave group-velocity spectra were defined via MFA (Chapter 2) considering an active acquisition (vertical-impact sledgehammer, offset 40 m (see seismic traces in Figure A8.2(a))), while H/V spectral ratio was computed thanks to a 20-min passive acquisition.

The results of the joint inversion of the ZVF group-velocity spectrum (processed following the FVS approach) and the HVSR curve are synthetically presented in

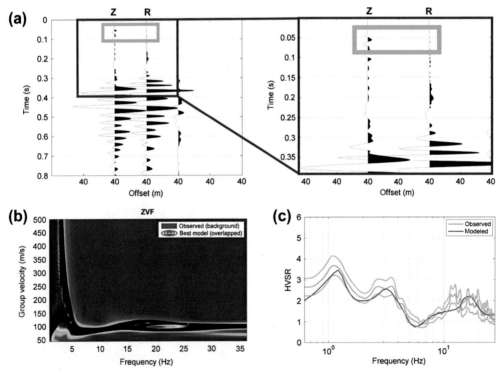

Figure A8.2 Site#2: results of the joint inversion of ZVF group-velocity spectrum and HVSR. In the upper panel the acquired seismic traces (active seismics) for group-velocity analysis via MFA (Z and R indicate the vertical and radial components, respectively). The green rectangle highlights the P-wave refraction related to the water table (1.8 m deep) (see text for details). In the lower panel: (b) the observed (background colors) and synthetic (overlaying contour lines) group-velocity spectra; (c) the observed and synthetic HVSR. Identified model is reported in Figure A8.3.

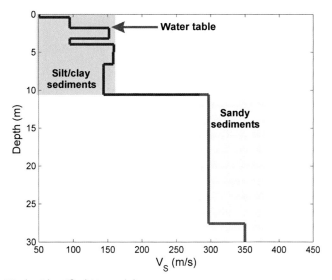

Figure A8.3 Site#2: the identified V_S model.

Figure A8.2: as for the rest of the book, background colors represent the field velocity spectrum while overlaying contour black lines the velocity spectrum of the identified model (reported in Figure A8.3). Please notice the very good agreement for both the HVSR curves and the observed and synthetic group-velocity spectra that almost perfectly overlap.

Also indicated the P-wave refraction (in the green rectangle) related to the water table (about 1.8 m deep) and which appears quite evident on the vertical component but definitely less clear in the radial one (the reason for that is clarified in the case study#2, where an analogous signal is reported and commented).

Cone Penetration Test and piezocone (CPTUs) performed down to a depth of 10 m, showed extremely low values for both q_c (cone resistance; average value around 1 MPa), f_s (local unit side friction resistance; average value around 0.02 MPa), and F_R (Friction Ratio; average value around 2.5%), in agreement with the low shear-wave velocities identified by the presented joint analysis down to about 11 m (Figure A8.3).

The remarkable V_S increase at that depth is easily interpreted as a stratigraphic passage from very soft sediments dominated by silt and clay (which below about 1.8 m are saturated (see refraction highlighted in Figure A8.2)) to a sequence where the sandy component significantly increases (we are nearby a lagoon area in North-East Italy and the deep sandy materials indicate a higher energetic regime related to the older shoreline).

Since the presented analyses indicate that the sands are not too deep, we might wonder whether the use of the HVSR was actually necessary to define that horizon and if, in this case, an approach based on just the active data (having considered only a single 3-component geophone we can define just the group velocities) could be sufficient.

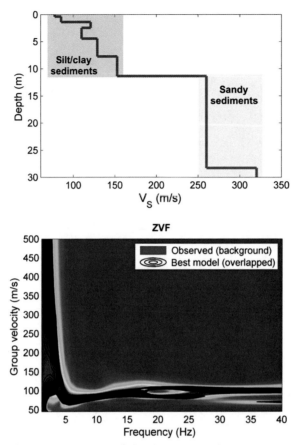

Figure A8.4 Results of the FVS inversion of the ZVF group-velocity spectrum. Background colors represent the field data, overlaying contour lines the synthetic data.

An FVS inversion of the ZVF group-velocity spectrum was then performed with the results reported in Figure A8.4 which shows that (due to the limited depth) the identification of the silt-to-sand passage is quite apparent (and substantially in agreement with the results of the above-illustrated joint inversion) also considering only the group-velocity spectrum obtained while considering a single trace (in this case the vertical component from an active acquisition (see trace Z in Figure A8.2)).

SITE#8

In a more northern area (site#8 in Figure A8.1) about 2.5 km from site#2, CPTU showed that the thickness of the silt/clay cover is slightly thinner (about 8.5 m (Figure A8.5)).

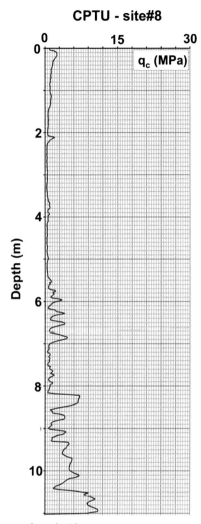

Figure A8.5 CPTU (cone resistance) at site#8.

For this site only active data were considered (vertical-impact sledgehammer, offset 40 m).

As highlighted in Chapter 2 (Paragraph 2.2.2), by considering a single (triggerable and calibrated) 3-component geophone is in fact actually possible to define four objective functions: three based on the group-velocity spectra of the ZVF, RVF (Rayleigh waves), and THF (Love waves) components and one based on the RVSR.

Since in this case we considered only a vertical-impact source (so Rayleigh waves only), a three-objective joint inversion was performed while considering the group-velocity spectra of the radial and vertical components of Rayleigh waves (FVS approach) jointly with RVSR (Figure A8.6 summarizes the results).

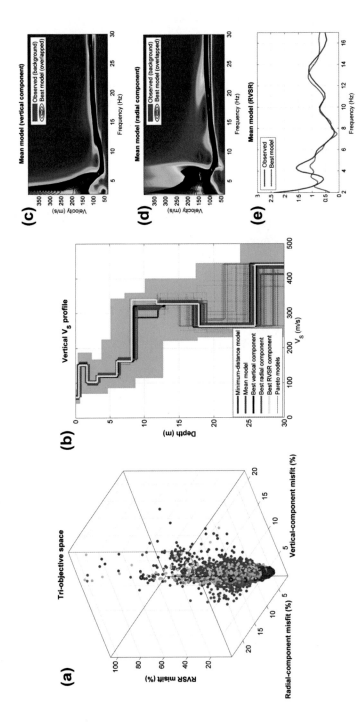

Figure A8.6 Site#8: results of the holistic (joint) inversion of the FVS inversion of the ZVF and RVF group-velocity spectra jointly with the RVSR (a patent-pending methodology): (a) model distribution in the three-objective space (reported the misfits for the three considered objective functions); (b) V_S profiles for the "best models" (green area represents the adopted search space); (c) and (d) the group-velocity spectra for the vertical and radial components (background colors represent the field data, overlaying contour lines the synthetic data); (e) observed and synthetic RVSR.

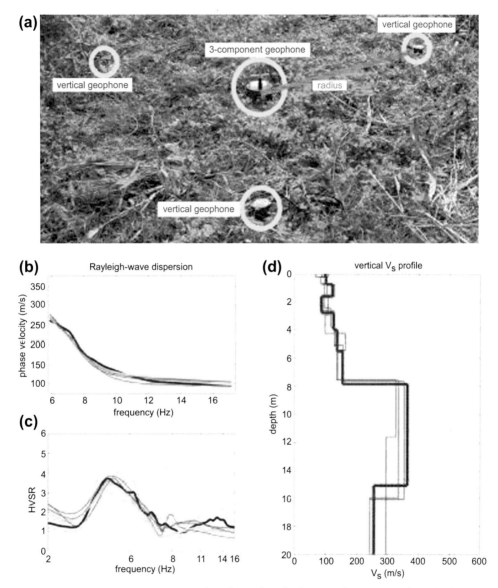

Figure A8.7 Site#8: joint acquisition and analysis of Rayleigh-wave dispersion and HVSR: a) acquisition geometry aimed at simultaneously defining Rayleigh-wave dispersion via MAAM and HVSR curve (three vertical geophones deployed in a 1.5m radius circular array and a central 3-component geophone); b) observed and synthetic *effective* (Tokimatsu et al., 1992) dispersion curves; c) observed and synthetic HVSR curves; d) identified V_S profile.

The pointy model distribution in the three-objective space (Figure A8.6(a)) demonstrates the overall consistency of the accomplished joint inversion (see Multi-objective Evolutionary Algorithms analyses in Chapter 5) while the overall agreement of the field and synthetic data is clear for both the velocity spectra (Figure A8.6(c) and (d)) and the RVSR (Figure A8.6(e)).

The accord with the CPTU data reported in Figure A8.5 is quite apparent.

Nearby the centre of such an active acquisition, a further (merely passive) procedure aimed at determining the V_S vertical profile was applied while considering Rayleigh-wave phase velocities defined via *Miniature Array Analysis of Microtremors* (MAAM - Cho et al, 2013) jointly inverted with HVSR.

This approach is quite interesting since it requires a light equipment and a very limited area (just a few meters), making such a technique quite relevant especially while working in urban areas.

In this case were used three 2 Hz vertical geophones defining an equilateral triangle inscribed in a circle with a radius of 1.5 m and a central 3-component 2 Hz geophone (with identical response curves) whose data were exploited both for the noise estimation (see Cho et al, 2013) and for defining the HVSR curve. In other words, by using just one 3-component geophone and three vertical-component geophones (Figure A8.7a), we can jointly define the Rayleigh-wave phase-velocity dispersion curve (via MAAM) and the H/V spectral ratio.

The acquisition geometry and the results of the joint inversion performed while considering the scheme described in Dal Moro (2010), are reported in Figure A8.7 and the consistency with the V_S profile reported in Figure A8.6 (in particular the identification of the top of the sandy layer) appears quite evident.

CASE STUDY 9

A Landslide Area

Focus: a joint analysis (vertical component of Rayleigh waves, horizontal-to-vertical spectral ratio (HVSR), and P-wave refraction) to quickly characterize a landslide area.

As often happens, the phase-velocity spectrum reported in Figure A9.1 shows a nontrivial energy distribution among different modes. The fundamental mode is in fact clear only in the 12–30 Hz frequency range (phase velocities between 270 and 160 m/s) while the dominating signal between about 5 and 13 Hz is instead related to higher modes (actually not to a single one but to the combination of several overtones). At lower frequencies, the energy appears quite low probably also because of the used source (a standard sledgehammer).

First arrivals (labeled as R) are clearly related to P-wave refraction.

For educational purposes, in order to put in evidence the almost-hidden fundamental mode, Figure A9.2 shows the effect of an *f-k* filtering (Yilmaz, 1987; Baker, 1999). For an alternative approach aimed at separating different modes, see Luo et al. (2009).

The HVSR curve reported in Figure A9.3 (data were acquired in the middle of the MASW array) shows two moderate lithological peaks (at 1.2 and 8 Hz) and one large industrial signal (at about 24 Hz) that we will clearly ignore.

The solution pursued via joint analysis of Rayleigh-wave dispersion (while adopting the Full Velocity Spectrum approach) and HVSR with the results reported in Figures A9.4 and A9.5.

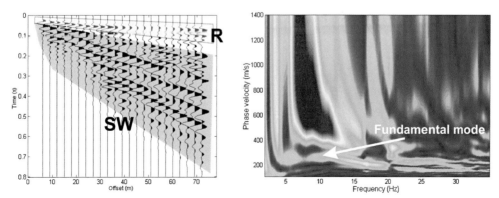

Figure A9.1 ZVF component (vertical geophones, vertical-impact source): seismic traces and phase-velocity spectrum. Areas labeled as R and SW indicate P-wave refraction event and Rayleigh wave, respectively. In the low frequency range (5–12 Hz) the energy associated to the fundamental mode appears remarkably lower than that related to higher overtones.

Surface Wave Analysis for Near Surface Applications
ISBN 978-0-12-800770-9, http://dx.doi.org/10.1016/B978-0-12-800770-9.15009-5

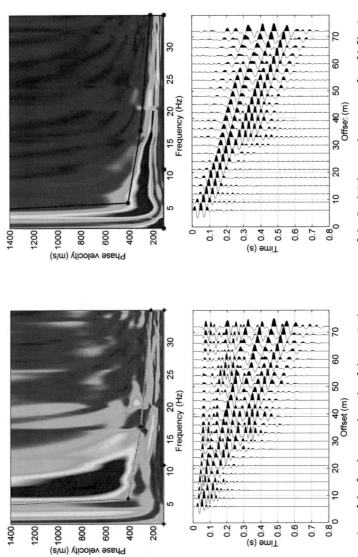

Figure A9.2 Separation of the fundamental mode of the vertical component of the Rayleigh wave by means of an *f-k* filtering: on the left column, the original data (the polygon reported on the velocity spectrum indicates the area/components that will be kept); on the right column, the filtered data now showing the fundamental mode previously somehow obscured by the energy related to higher overtones.

Industrial peak

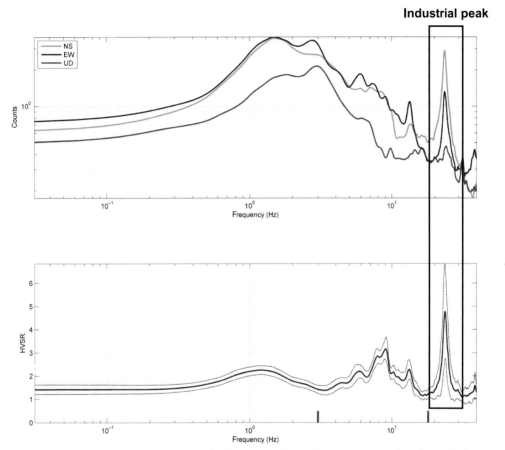

Figure A9.3 Observed amplitude spectra for the NS (North-South), EW (East-West) and UD (Up-Down, i.e. vertical) components, and horizontal-to-vertical spectral ratio (HVSR). The rectangle centered at about 24Hz highlights an "industrial" signal (i.e., related to some industrial activity) while the smoother HVSR peaks at about 1.2 and 8 Hz are related to lithological features (see Chapter 4).

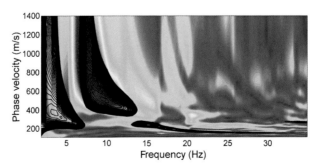

Figure A9.4 *Multichannel Analysis of Surface Waves* data inverted via Full Velocity Spectrum analysis. Background colors report the phase-velocity spectrum of the field data (same as Figure A9.1) while the overlaying black contour lines refer to the velocity spectrum of the identified model (see Figure A9.5). It must be underlined that the energy in the 6–14 Hz range is not related to a single higher mode but it represents the combined effect of several (approximately five) higher modes (see also Tokimatsu et al., 1992).

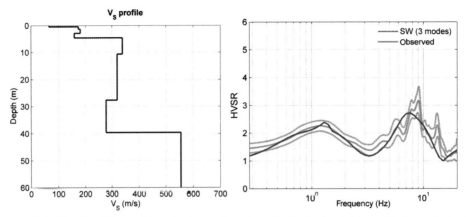

Figure A9.5 Identified V_S model and observed and synthetic horizontal-to-vertical spectral ratio (HVSR).

Eventually, since surface-wave dispersion and HVSR are not significantly influenced by P-wave velocities (see Chapters 1 and 3), we then quickly modeled the V_P values of the shallowest layers in order to match the early arrivals related to the refraction event (see Figure A9.6).

The P-wave refraction suggests that, at a depth of about 3 m, the water content in the shallow porous sediments remarkably increases (V_S around 150–180 m/s and V_P about 1700 m/s—see also case study 2) probably creating some critical condition in the slope instability.

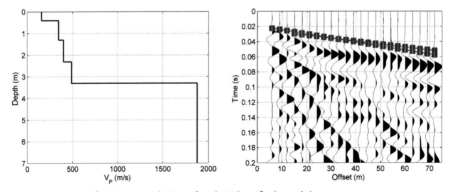

Figure A9.6 P-wave refraction travel times for the identified model.

CASE STUDY 10

Back to the Swiss Alps

Focus: a further example of multicomponent analysis of Rayleigh (radial and vertical components) and Love waves acquired (once again) in the Swiss Alps and analyzed jointly with body-wave refraction travel times.

Similarly to the case study 6, this *dataset* was acquired in the framework of a series of surveys managed by the *Swiss Seismological Survey* and performed by the author and Lorenz Keller (*roXplore*, Switzerland). Due to the high standards required by the client, in order to acquire all the components relevant for surface-wave analysis (ZVF, RVF, THF), during the acquisition campaign we used both vertical (4.5 Hz) and horizontal (10 Hz) geophones.

Figure A10.1 reports (from top to bottom) field traces of the ZVF, RVF, and THF components, their phase-velocity spectra and the phase-velocity spectra of the model identified via joint *Full Velocity Spectra* inversion (see Figure A10.2).

Three short comments about the data reported in Figure A10.1:

1. The velocity spectra of the vertical and radial components of Rayleigh waves are different (this is clearly normal since the respective Green's functions are different and consequently, in general terms, the velocity spectra are different—how much it depends on the site).

2. The agreement between the observed phase-velocity spectra (middle panel) and the ones of the identified model (lower panel) is quite apparent (see also Figure A10.3 where the spectra are shown overlaid).

3. The three datasets (ZVF, RVF, and THF components) used in this case can be thought as the three polygons constraining the solution in Figure 5.8 where the concept of joint analysis (i.e., inversion) is schematically represented.

Some comments about the P-wave refraction and its manifestation on phase-velocity spectra.

In the upper left plot of Figure A10.1 (where ZVF traces are reported) a rectangle (marked by the letter R) puts in evidence the P-wave refraction events that are much less evident in the RVF component, where—in order to highlight the P-wave refraction—some gain is necessary (see Figure A10.4 and related text).

In the ZVF phase-velocity spectrum of the observed traces reported in Figure A10.1 (middle panel), a rectangle (indicated by the letter R) in the upper right corner indicates the energy related to the P-wave refraction (high frequency and high velocity). In order to make it even clearer, in the Figure A10.6 a zoomed version of these data is reported.

Surface Wave Analysis for Near Surface Applications
ISBN 978-0-12-800770-9, http://dx.doi.org/10.1016/B978-0-12-800770-9.15010-1

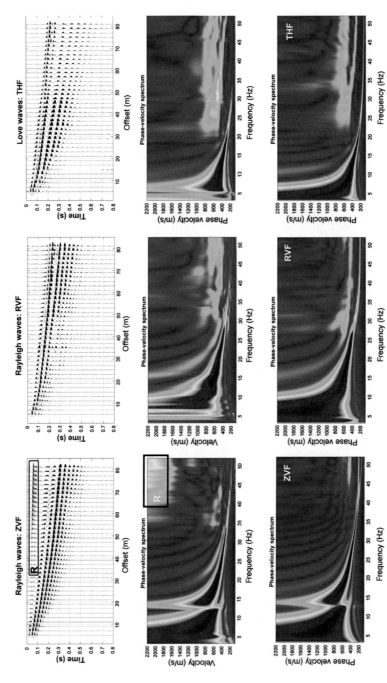

Figure A10.1 Top panel (from left to right): field traces of the ZVF, RVF, and THF components; middle panel: phase-velocity spectra of the field data; bottom panel: phase-velocity spectra of the model identified via joint Full Velocity Spectrum inversion.

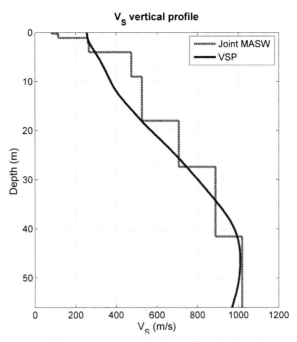

Figure A10.2 Comparison of the shear-wave velocity profile obtained from *vertical seismic profile* (VSP) and the one identified from the multicomponent joint analysis of surface waves via Full Velocity Spectrum inversion. Numbers report the V_P values set while optimizing the match with the P-wave refraction travel times (see also Figures A10.4 and A10.5). MASW, multichannel analysis of surface waves.

Figure A10.4 shows the ZVF, RVF, and THF traces with, overlaying, the computed refraction travel times for the identified model (Figure A10.2 and A10.5). In order to focus our attention of just the first arrivals, only the first 0.25 s are shown and an *automatic gain control* (AGC) is applied. The rectangle in Figure A10.4(b) highlights the radial component of the P-wave refraction that, due to its amplitude (which depends on the critical angle), is quite weak in the nonamplified traces (see Figure A10.1). On the other hand, within the triangle in Figure A10.4(c) it is shown some incoherent noise whose amplitude was increased by the application of the AGC (compare with the THF traces reported in Figure A10.1).

Eventually, Figure A10.5 reports the zoomed version of the Figure A10.4(a) together with the V_P profile of the shallow layers. Such V_P model is obtained from the one reported in Figure A10.2 while adopting the appropriate Poisson ratios (the reader must remember that surface-wave propagation and dispersion are not significantly influenced by the V_P values, so, in case P-wave velocities are required), it is necessary to consider the P-wave refraction travel times via simple forward modeling (as in the present case) or by means of more sophisticated analyses (see, e.g., Dal Moro, 2008).

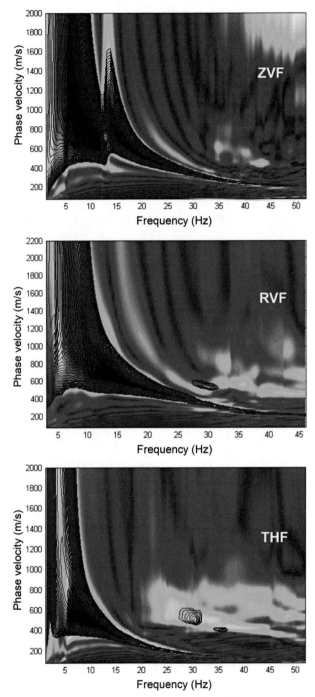

Figure A10.3 Observed (background colors) and inverted (overlaying contour lines) velocity spectra for the ZVF, RVF, and THF components (same spectra reported separately in Figure A10.1).

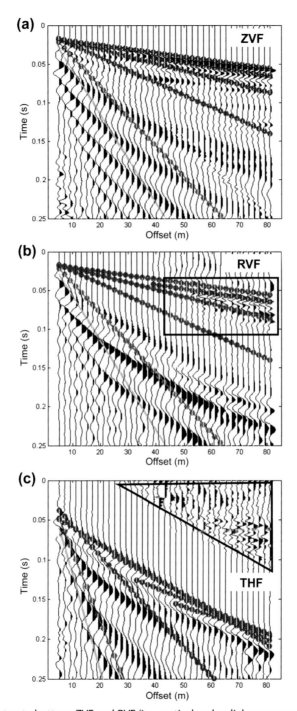

Figure A10.4 From top to bottom: ZVF and RVF (i.e., vertical and radial components, respectively) and THF traces after the application of an *automatic gain control*. First arrivals on the first two datasets relate to P-wave refraction while the first arrivals on the THF component relate to SH-wave refraction. Red lines relate to the refractions generated by all the horizons in the model (see Figure A10.2) but clearly only the very first arrivals are so-to-say visible. Green lines relate the direct wave. In the (a) and (b) graphs the P-wave refraction travel times are reported, while the SH-wave refraction travel times are reported in the (c) plot.

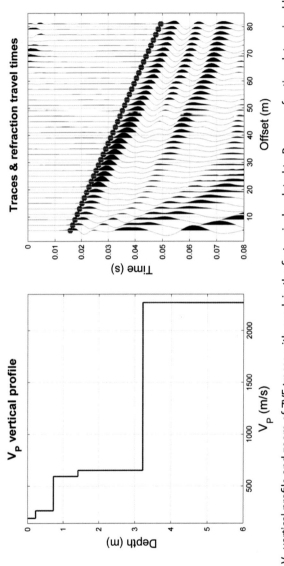

Figure A10.5 V_P vertical profile and zoom of ZVF traces with overlain the first arrivals related to P-wave refraction determined by the adopted V_P model.

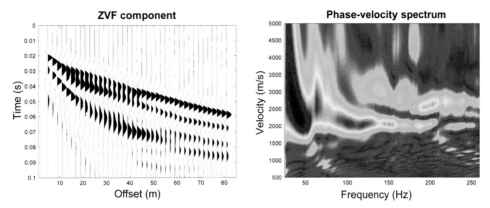

Figure A10.6 On the left the first arrivals (P-wave refraction with possibly some guided waves in it as well) recorded by the ZVF traces; on the right, the phase-velocity spectrum computed for high velocities and high frequencies and showing up the velocities related to these early (P-wave refraction) events.

By comparing the V_S and V_P velocities in the shallowest area it is possible to put forward the idea that the materials (basically related to glacial sediments dominating the area) below the depth of about 3.5 m are saturated. In fact, in case of soft sediments, a high water content produces a large difference in the V_P/V_S ratio (thus in the Poisson's ratio): water will indeed significantly affect (increase) the velocity of the compression waves while will have minor effects on the shear-wave velocity (see also case study 2 and Paragraph 7.2.2).

ACKNOWLEDGMENTS

The author wishes to thank Lorenz Keller (*roXplore*) for his friendship (and continuous effort) and *Nagra (Nationale Genossenschaft für die Lagerung radioaktiver Abfälle)* for the permission of publishing the data.

CASE STUDY 11

Modes and Components (A Very Tricky Site)

Focus: a noticeable example of complex Rayleigh-wave phase-velocity spectrum that would quite hard to solve by adopting an approach based on the identification of the *modal* dispersion curves and without also considering Love waves. Solution is in fact sought following the Full Velocity Spectrum (FVS) approach (an extension of the apparent dispersion curve resulting from mode superposition and described by Tokimatsu et al. (1992)). To further constrain the analysis and thus also further validate the identified model, horizontal-to-vertical spectral ratio (HVSR) curve is also considered.

In order to depict the local V_S vertical profile (as we will see, this site will reveal one of the trickiest of the whole book), three datasets were acquired and considered: ZVF (vertical component of Rayleigh waves), THF (Love waves), and HVSR.

Seismic traces and phase-velocity spectra reported in Figure A11.1 make immediately clear a very common fact: Love-wave velocity spectrum appears much *simpler* than the one of Rayleigh waves.

HVSR analyses are reported in Figures A11.2—A11.4: a remarkable peak verifies at about 16 Hz (and incidentally meets five out of six SESAME *criteria for a clear H/V peak*) while in the 1.8—5 Hz frequency range, the H/V ratio results smaller than 1. The overall good permanence (continuity over the time—Figure A11.3) and the absence of relevant directivity (Figure A11.4) suggest that the obtained H/V curve is sufficiently representative of the site.

As we will fully understand only after having evaluated the analyses, in this case the possibility of considering Love waves and HVSR revealed essential because the ZVF component alone would otherwise be completely misleading being this probably the most complex dataset presented.

A three-component joint inversion via multiobjective evolutionary algorithms (see Chapter 5) was performed while considering the dispersion of Rayleigh and Love waves (in the FVS perspective described in Chapter 6 and adopted in a number of presented case studies) jointly with the HVSR curve. Results are summarized in Figures A11.5—A11.7 which are also providing the evidence of the overall good consistency of the identified solution: phase-velocity spectra of the field datasets and of the identified model are in fact quite consistent (they entirely overlap—see Figure A11.5) as well as the observed and synthetic HVSR curves (Figure A11.6).

Surface Wave Analysis for Near Surface Applications
ISBN 978-0-12-800770-9, http://dx.doi.org/10.1016/B978-0-12-800770-9.15011-3

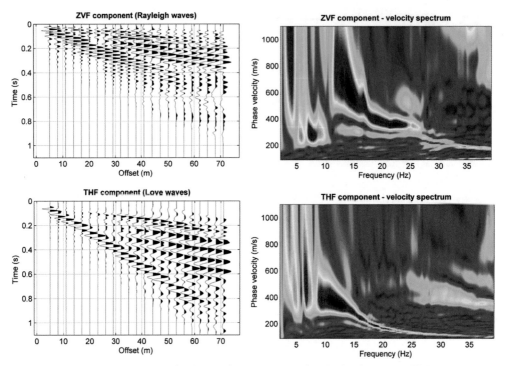

Figure A11.1 Seismic traces and phase-velocity spectra for Rayleigh waves (ZVF component reported in the upper panel) and Love waves (THF component reported in the lower panel).

It should be underlined that the low-velocity layers in the identified V_S model (Figure A11.7) are responsible for both the complex velocity spectra and for the H/V values smaller than 1 in the 1.8–5Hz frequency range (see HVSR curves in Figure A11.6).

If we compare the data reported in Figure A11.5 and the modal dispersion curves in Figure A11.8, the extremely complex relationships between the *modal* dispersion curves and the *effective* dispersion curves result quite apparent.

The presented analyses show some relevant facts, also put in wide evidence in Chapter 3 (and in the therein mentioned literature), that can be summarized in the following schematic points:

1. Especially while considering Rayleigh waves, the fundamental mode does not necessarily dominate the data.
2. The continuity of a signal in the velocity spectrum does not necessarily mean that a single mode is responsible for that feature.
3. The signals dominating the ZVF velocity spectrum cannot be described in terms of modal dispersion curves because, especially when multiple V_S inversions take place, the apparent dispersion curve can be the result of the energy contribution of several modes: signals A and B in Figure A11.8 are the result of several higher modes. In fact,

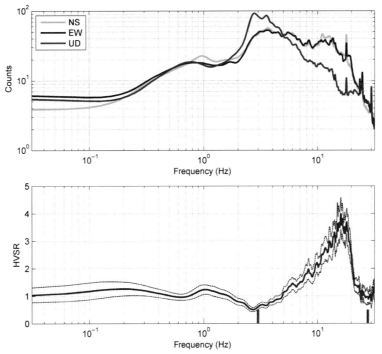

Figure A11.2 Amplitude spectra and horizontal-to-vertical spectral ratio (HVSR). Please notice the large peak at about 16Hz, the smaller one around 1Hz, and the values smaller than 1 between 1.8 and 5Hz. NS, North-South; EW, East-West; UD, Up-Down (i.e., vertical).

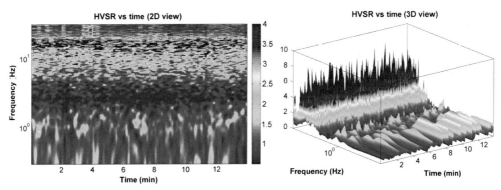

Figure A11.3 Horizontal-to-vertical spectral ratio (HVSR) continuity over the time (window length l_{w}, 16 s).

signal A is fundamentally related to the 8th to 15th higher modes (this is the reason why it spans in a wide velocity range, from about 600 to 1000 m/s) while signal B is largely dominated by the third higher mode (compare with synthetic velocity spectra reported in Figure A11.5).

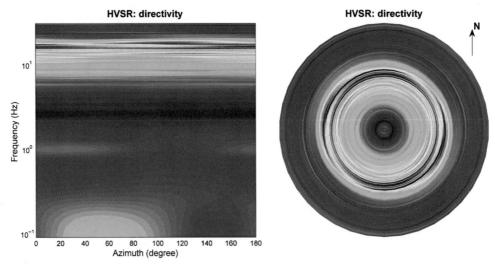

Figure A11.4 Directivity of the H/V spectral ratio: HVSR curves are represented as a function of the angle formed with the NS direction (please notice that the 16 Hz peak shows no dependence on the azimuth).

Figure A11.5 Results of the joint analysis performed while considering Rayleigh- and Love-wave dispersion (processed according to the FVS approach) together with the H/V spectral ratio (see Figure A11.6). Upper and lower panels report the ZVF and THF components, respectively. Background colors represent the field data while the overlaying contour lines represent the synthetic velocity spectra of the model reported in Figure A11.7.

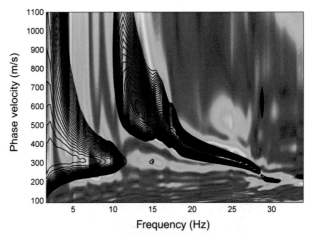

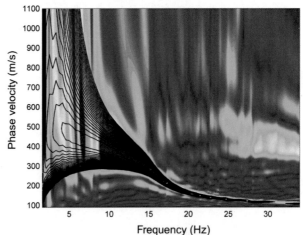

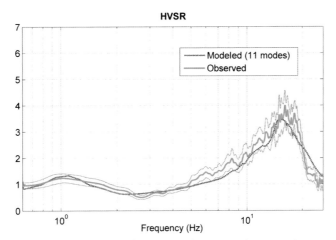

Figure A11.6 Horizontal-to-vertical spectral ratio (HVSR): green and magenta lines report the observed and synthetic H/V spectral ratios. Synthetic curve is computed according to Arai and Tokimatsu (2004) while considering the model shown in Figure A11.7.

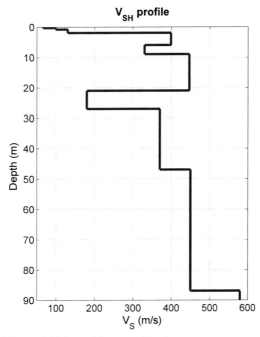

Figure A11.7 Identified V_S model (shown down to 90 m).

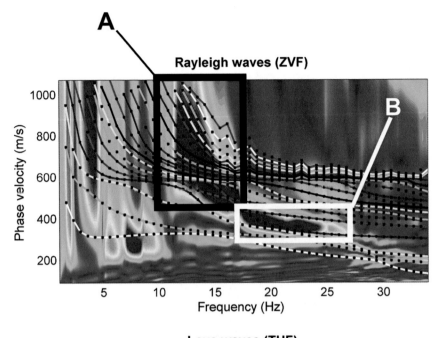

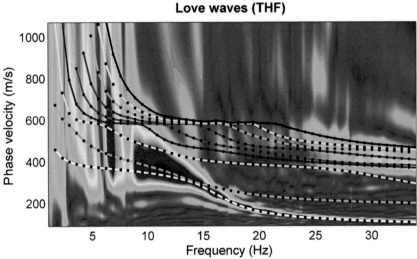

Figure A11.8 Field-velocity spectra for Rayleigh and Love waves with, overlaying, the modal dispersion curves for the V_S model reported in Figure A11.7. The complex relationships between modal and *effective* dispersion curves (i.e., the velocity spectra) are apparent (see synthetic velocity spectra in Figure A11.5 and Chapter 3). Signals A and B are commented in the text.

4. When multiple velocity inversions occur (see model reported in Figure A11.7), also the velocity spectra of Love waves (notoriously simpler in their propagation and dispersion) can result not straightforward in their meaning: in this case (compare Figures A11.5 and A11.8), the wide phase velocity range characterizing the velocity spectrum between about 5 and 12 Hz (in this frequency range Love wave propagates with velocities from about 350 to 550 m/s) is due to the joint effect of the first three modes (which cannot be singularly resolved).

5. In cases like this, only a joint analysis capable of properly considering a sufficient amount of data (Rayleigh and Love waves plus, for imaging the deepest part, the H/V spectral ratio) can actually solve the site and provide an accurate V_S profile (almost useless to say that traditional analyses based on the *modal* dispersion curves would result impracticable).

CASE STUDY 12

Analyzing Phase and Group Velocities Jointly with Horizontal-to-Vertical Spectral Ratio

Focus: a comprehensive study of a site while considering a set of active and passive acquisitions (horizontal–to-vertical spectral ratio, (HVSR), Extended Spatial Autocorrelation (ESAC), and Love-wave group velocities from an active acquisition performed while considering a single horizontal geophone).

The results of a joint inversion performed while considering HVSR and group-velocity spectrum of Love waves are compared with the apparent dispersion curve obtained from an ESAC acquisition, also putting in evidence that this latter is actually the result of two modes.

A series of acquisitions were performed with the aim of acquiring three datasets aimed at performing a comprehensive and comparative analysis. The considered datasets are

1. HVSR (Figure A12.1),
2. Love-wave group velocities computed via *Multiple Filter Analysis* (MFA) while considering a single trace (Figure A12.2), and
3. Rayleigh-wave phase velocities obtained via ESAC.

A12.1 JOINT INVERSION OF HVSR AND LOVE-WAVE GROUP VELOCITIES

A joint inversion of the HVSR curve and the THF group-velocity spectrum according to the Full Velocity Spectrum approach (see Chapter 6) was performed following a biobjective procedure (see Chapter 5) with the results summarized in Figure A12.3.

A12.2 ESAC DATA

In order to compare the results obtained following the above-presented procedure, an ESAC analysis (Rayleigh-wave phase velocities) was also considered. Figure A12.4 reports the first 6 min of the considered dataset and the channel map showing the position of the 22 vertical-component geophones (logistical aspects prevented from disposing the geophones on a larger area).

Surface Wave Analysis for Near Surface Applications
ISBN 978-0-12-800770-9, http://dx.doi.org/10.1016/B978-0-12-800770-9.15012-5

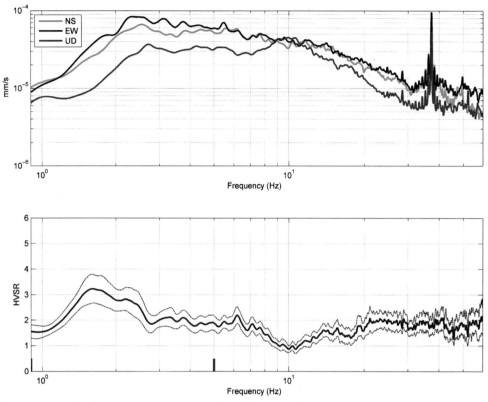

Figure A12.1 Amplitude spectra (upper panel) and horizontal-to-vertical spectral ratio (HVSR) (lower panel) for the site under study. In the amplitude spectra, at about 37 Hz, there are some clear industrial peaks (see Chapter 4) that, because the amplitude is the same on both the horizontal and vertical components, do not produce a related peak on the H/V curve. Since frequencies higher than 20 Hz relate to very shallow features, they were not considered in the following analysis. NS, North-South; EW, East-West; UD, Up-Down (i.e., vertical).

In Figure A12.5 are reported the ESAC velocity spectrum (background colors) and, overlaying, the *modal* and the *effective* dispersion curves of the model shown in Figure A12.3(c) (identified by the joint THF + HVSR inversion briefly described in the previous section).

The comparison of the results allows putting in evidence a couple of remarkable points:

1. The THF (group velocity) + HVSR analysis results in very good agreement with the ESAC data.

2. In spite of its apparent continuity, the ESAC *effective* dispersion curve is actually the result of two modes (the fundamental and the first higher mode), and that means that (like any other method that can be used to determine the dispersive properties of a medium) it necessarily contains some ambiguity (see Chapters 1 and 3).

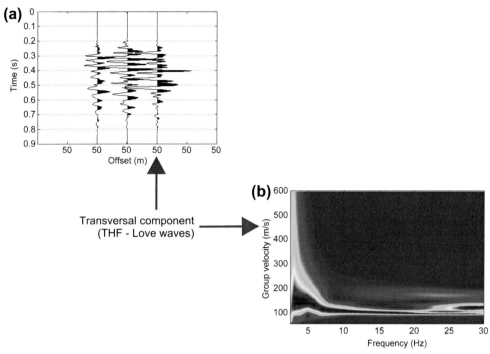

Figure A12.2 Active data considered to determine Love-wave group velocities: (a) the transversal component (third trace) acquired while considering a shear source (offset 50 m) and (b) its group-velocity spectrum.

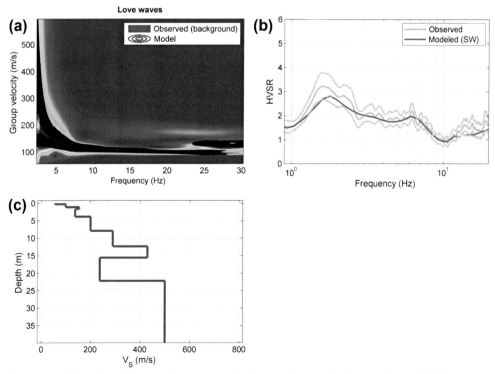

Figure A12.3 Results of the joint inversion of the horizontal-to-vertical spectral ratio (HVSR) and the THF group-velocity spectrum: (a) background colors represent the group-velocity spectrum of the field trace while overlaying contour lines relate to the velocity spectrum of the identified model, (b) observed and synthetic HVSR, (c) V_S profile from the joint inversion.

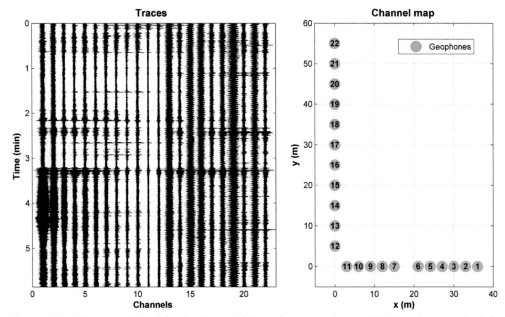

Figure A12.4 Data considered for the Extended Spatial Autocorrelation (ESAC) analysis: on the left, the first 6 min of the considered dataset and on the right, the channel map.

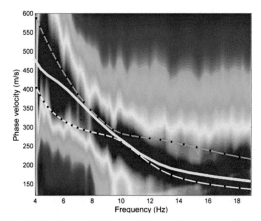

Figure A12.5 Extended Spatial Autocorrelation and dispersion curves. The background colors represent the phase-velocity spectrum obtained via ESAC analysis; the white and green dotted lines are the modal dispersion curves of the first two modes of the model reported in Figure A12.3(c) and the yellow thin line is the respective *effective* (or *apparent*) dispersion curve.

Using only the ESAC approach, the obtained Rayleigh-wave effective dispersion curve would then be potentially dangerous if we would decide to interpret that curve as pertaining to the fundamental mode. Reader must carefully consider that an *apparent* dispersion curve having that same trend but dominated uniquely by the fundamental mode necessarily also exists. In other words: the same *apparent* dispersion curve can refer to different models that excite different modes that anyway combine in such a way to create a continuous *apparent* dispersion curve. As widely discussed in this book, only the joint analysis of several datasets is capable of solving uncertainties and ambiguities (see Chapter 5).

The acquisition procedures and the equipment necessary to perform the considered acquisitions (MFA, HVSR, and ESAC) are anyway quite different. For the very first joint analysis (group-velocity spectrum obtained via MFA analysis and HVSR), it is only necessary to have a calibrated 3-component geophone which is also capable of acquiring active data (i.e., triggerable). On the second case, a multichannel seismograph and a sufficiently large number of geophones are necessary.

CASE STUDY 13

Some Focus on Horizontal-to-Vertical Spectral Ratio Computation

Focus: while assessing horizontal-to-vertical spectral ratio (HVSR) curves (see Chapter 4), it is important to properly consider some computational issues that can otherwise spoil the analyses. This is particularly important when possible H/V peak(s) undergo analyses aimed at defining their statistical robustness, and thus their validity especially with respect to seismic hazard studies. SESAME criteria (AA.VV., 2005) are a set of statistical values that help in evaluating the stability of an H/V peak, but their computation requires special care especially when dealing with multipeak HVSR curves.

A13.1 SPECTRAL SMOOTHING

As reported in Chapter 4, the computation of the amplitude spectra of the three considered components (North-South, East-West and Vertical) is the key to successively obtain the H/V curve and the *noise* (in the broad sense) necessarily present in the seismic traces will clearly mirror in the amplitude spectra as well (and, consequently, in the HVSR curve).

After having removed transient events from the data (see Chapter 4) for computing the H/V curve, it is necessary to fix the values of some key parameters such as the *tapering*, the *window length*, and the *smoothing* percentage. While for the first two parameters the reader is referred to the literature mentioned in Chapter 4, here we briefly focus on the role of the *spectral smoothing*.

In fact, any mathematical operation performed on the data (including the *smoothing* of the amplitude spectra), necessarily alters them and we must then carefully consider its effects, which can otherwise go beyond our intentions.

Fundamentally, there are two sets of SESAME criteria (AA.VV., 2005): the first three ("*criteria for a reliable H/V curve*") refer to the curve as a whole (they help you in understanding whether the collected data are qualitatively and quantitatively sufficient to define the H/V curve) while a second set of statistical values (the six "*criteria for a clear H/V peak*") helps assessing the statistical robustness of a peak.

Of course, such criteria must be considered in a flexible perspective and not as a sort of rigid and sclerotic framework. It is then clear that, in order to properly consider them, we must be aware of some critical aspects.

In Figure A13.1 are shown the amplitude spectra (computed while considering a 2% spectral smoothing) for the three considered components (two perpendicular horizontal

Surface Wave Analysis for Near Surface Applications
ISBN 978-0-12-800770-9, http://dx.doi.org/10.1016/B978-0-12-800770-9.15013-7

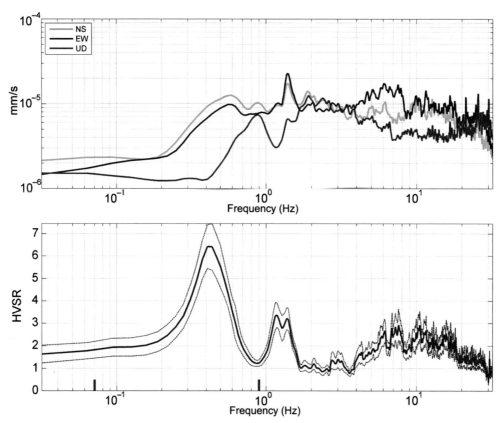

Figure A13.1 Amplitude spectra (upper panel) and horizontal-to-vertical spectral ratio (HVSR) (lower panel) for a site nearby the Grado-Marano lagoon (NE Italy). *Spectral smoothing* (2%) is applied while considering a triangular window. NS, North-South; EW, East-West, UD, Up-Down (i.e., vertical).

sensors and one vertical) and the respective H/V curve for a dataset acquired in NE Italy in an area where a soft-sediment cover lies over a *bedrock* at a depth of few hundreds of meters. The very-well defined H/V peak at about 0.4 Hz results quite apparent (its value is around 6 and SESAME criteria are met, so that we can affirm its statistical robustness).

On the other side, by comparing such H/V curve with the one reported in Figure A13.2 (obtained while considering a 15% *spectral smoothing*), it is clear that the 0.4 Hz peak is now seriously altered: its position (0.3 Hz) and amplitude (3.7) are in fact quite different with respect to the curve presented in the previous Figure and its trend is now such that only three out of six SESAME criteria are met.

Moral: be careful when choosing the amount of smoothing especially when you are interested in very-low frequency peaks which, compared to high-frequency peaks, are more affected by excessive spectral smoothing.

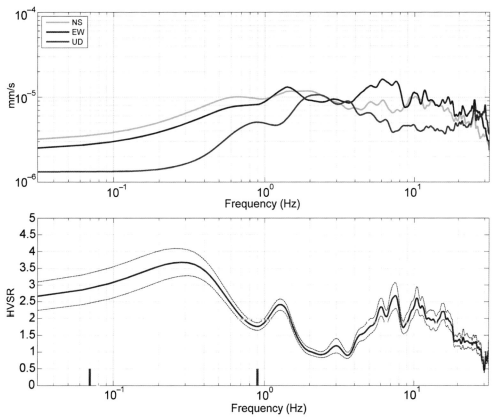

Figure A13.2 Amplitude spectra (upper panel) and horizontal-to-vertical spectral ratio (HVSR) (lower panel) for the same dataset considered in Figure A13.1, but now considering a *spectral smoothing* of 15%. NS, North-South; EW, East-West, UD, Up-Down (i.e., vertical).

Although it is impossible to indicate optimal and universal values, in general terms it can be suggested a limited smoothing (2–5%) for low-frequency peaks and a higher smoothing (up to 10–15%) for high-frequency H/V peaks.

A13.2 SESAME CRITERIA FOR MULTIPEAK HVSR CURVES

Since the second set of SESAME criteria refers to a specific peak (the six "*criteria for a clear H/V peak*"), it should be clear that, in case of multipeak HVSR curves, their computation should be performed separately for each of them.

To clarify this point we will now consider (Figure A13.3) an H/V curve having two distinct peaks (from the engineering point of view, frequencies higher than 20 Hz are completely irrelevant and are therefore ignored).

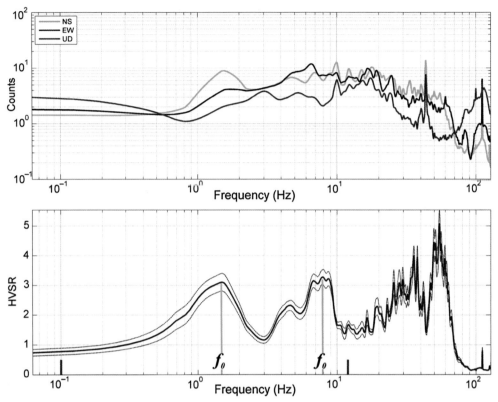

Figure A13.3 Amplitude spectra (upper panel) and horizontal-to-vertical spectral ratio (HVSR) (lower panel) for a site where the stratigraphic column produces two distinct peaks, one at 1.6 Hz and the other at 8 Hz (from the engineering point of view, the peaks at frequencies higher than 20 Hz are irrelevant and are therefore ignored). UD, up–down.

If we compute the six "*criteria for a clear H/V peak*" while considering the 0.1–12 Hz frequency range (see the two small red lines along the x-axis in Figure A13.3, lower panel), thus including both the peaks, we obtain the data summarized in Figure A13.4 where are reported all the HVSR curves computed for all the considered windows and the peak frequency for each of them. These two graphs put in evidence the fact that for some windows the largest peak is the one centered around 1.6 Hz, while for some other windows the largest peak is the one around 8 Hz.

It is clear that if we compute the SESAME statistical analyses while considering this bimodal distribution, the obtained parameters representing the stability of the peak (the six *criteria for a clear H/V peak*) will indicate very poor permanence and no *clear H/V peak* that meets the SESAME criteria will be declared.

However, as a matter of fact, such values (computed while considering the entire 0.1–12 Hz frequency range) are meaningless, because in order to avoid this sort of problems related to multipeak curves, those six parameters should be computed *peak by peak*.

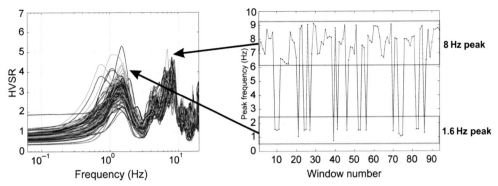

Figure A13.4 On the left, the horizontal-to-vertical spectral ratio (HVSR) curves computed for each considered windows (the dataset is divided into a number of small windows—see Chapter 4); on the right, the peak frequencies computed within each single window while considering the 0.1—12 Hz frequency range. Within some windows the dominating (highest) peak is the one with 1.6 Hz while in others is the one with 8 Hz (see also Figure A13.3).

In fact, if we limit the statistical analyses only around each single peak (in the $f_0/4-4f_0$ frequency range considered by the first two criteria—see AA.VV., 2005), the results are going to be purely representative of the stability of the considered peak, not being spoiled by the effects of further peaks.

In Figure A13.5, we report some fundamental parameters regarding the SESAME analyses performed to evaluate the stability of the 1.6 Hz peak while considering only the data in the 0.1—3 Hz frequency range. It should be clear that now the criteria related to the peak stability (the six *criteria for a clear H/V peak*) are not any longer spoiled by the bimodality of the curve and the 1.6 Hz peak can be finally (and properly) defined as significant since it meets at least five of the six criteria (see following box).

SESAME Criteria for a Clear H/V Peak (At least Five Should Be Fulfilled—See Figure A13.5 and, for further technical details, AA.VV., 2005)

Focus on the 1.6 Hz (f_0) peak, statistical analyses performed while considering the 0.1–3.0 Hz frequency range.

1. [exists f − in the range $[f_0/4, f_0]|A_{H/V}(f-) < A0/2$]: yes, at frequency 0.4 Hz (*OK*, the peak decreases to A/2 at 0.4 Hz, that is between f_0 and $f_0/4$).

2. [exists f + in the range $[f_0, 4f_0]|A_{H/V}(f+) < A0/2$]: yes, at frequency 2.4 Hz (*OK*, the peak decreases to A/2 at 2.4 Hz, that is, between f_0 and $4f_0$).

3. [A0 > 2]: 3.1 > 2 (*OK*, the peak is larger than 2).

4. [fpeak[$A_{H/V}$ (f) ± sigmaA(f)] = f_0 ± 5%]: (*OK*, the peak tolerably oscillates around the average point).

5. [sigmaf < epsilon(f_0)]: 0.223 > 0.156 (*NO*, a little too much variability in the f_0 computed for each window).

6. [sigmaA(f_0) < theta(f_0)]: 0.295 < 1.78 (*OK*, the variability of the peak amplitude is acceptable).

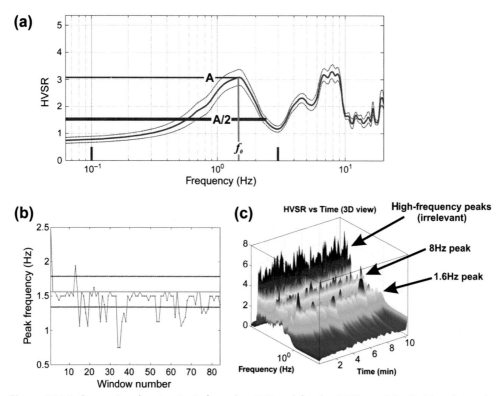

Figure A13.5 Computing the six *criteria for a clear H/V peak* for the 1.6 Hz peak by limiting the statistical analysis in the 0.1–3 Hz frequency range, thus avoiding the problems due to the bimodality of the horizontal-to-vertical spectral ratio (HVSR) curve (compare with the data reported in Figures A13.3 and A13.4): (a) fundamental parameters considered in the statistical analyses (peak frequency f_0, its amplitude A and half of it); (b) peak frequency for all HVSR curves computed for all the considered windows (considering only the data in the 0.1–3 Hz frequency range); (c) permanence of the HVSR curves over the time.

If we would like to assess the stability of the 8 Hz peak, we should then focus our SESAME analyses while considering only the data more or less in the 3–11 Hz frequency range (since the analyses would simply trace the same procedure used for the 1.6 Hz peak, presentation is omitted).

A couple of final remarks about the so-called *dominating* (or *fundamental—* unfortunately some terms are often used in a so-to-speak subjective way) HVSR peak.

In seismic hazard studies, very often the structural engineers require the *dominating* (or *fundamental*) resonance frequency and, sometimes, the *amplification factor*.

First of all, it must be underlined that the HVSR curve does not represent the site amplification, but rather a sort of estimation of the resonance frequency. In other terms, while the HVSR peak frequency (or frequencies) more or less properly identifies the local resonance frequency, the actual amplification is something to be computed on the basis of the local V_S profile and considering the earthquake representative of the local seismicity (e.g., Wen et al., 1994).

But, what would be the "dominating" frequency in cases similar to the one reported in Figure A13.3, that is, in cases where multiple peaks are present?

A universal answer actually does not exist and the geophysicist should report about all the HVSR peaks (assessing the "SESAME criteria for a clear H/V peak" for each of them). In fact, the HVSR peak, which is potentially dangerous from the structural point of view (these studies are usually performed to assess the seismic hazard with respect to some specific building or infrastructure), depends on the specific characteristic of the construction itself. Roughly speaking, it must be avoided that the site resonance frequency coincides with the resonance frequency of the construction (any structure has its own *eigenfrequency*).

As a consequence, referring, for instance, to the multipeak HVSR curve presented in Figure A13.3, if they plan to construct a two-storey house, the peak potentially dangerous is the one at 8 Hz, while if they plan to build a seven-storey building, the 8 Hz peak is quite irrelevant while the 1.6 Hz resonance can be a problem (in general terms, the taller the structure, the lower is its *eigenfrequency*). Of course, this sort of considerations must be quantitatively defined by the structural engineer (the *eigenfrequency* of a building does not depend only on its height, but on the used materials as well) and not by the geophysicist, who should just report the geophysical evidences regarding all the potential resonance frequencies and, if required, assess the ground motion in case of seismic event (e.g., Hashash and Park, 2001).

CASE STUDY 14

Surface Waves on the Moon

Focus: joint analysis of Rayleigh-wave dispersion (group velocities determined via active seismics) and *horizontal-to-vertical spectral ratio* (HVSR) (determined by means of 3-component recording of a meteoric impact) aimed at identifying the thickness of the Lunar regolith and the shallow shear–wave velocity profile. The peculiar geomechanical characteristics of the Lunar materials (which are clearly related to specific formation processes) mirror in very low (*extraterrestrial*) V_S values accompanied to unusually high Q_S quality factors.

A14.1 A BRIEF COMPULSORY FORWARD

The *Apollo program* probably represents one of the highest achievements of the modern times not only with respect to its technological and scientific outcomes, but most of all for its incommensurable epical meaning. In this regard, the Apollo legacy projects the Lunar exploration and all the related events closer to the ancient Greek mythology rather than in the realm of the mere technological progress (the tragic Apollo 1 cabin fire that killed the three crew members, the dramatic Apollo 13 incident and the consequent return to the Earth, the millions of people which spontaneously gathered around the *Lyndon Johnson Space Center* in Houston to attend the launches and the countless number of people that over the years watched the pictures and the footage realized by the astronauts, etc.).

All this, together with the scientific data that were collected and analyzed, undoubtedly embodies an invaluable world heritage for the whole mankind.

With this final case study, the author also intends to express his sincere and profound gratitude to Prof. Bob Kovach (*Stanford University*), who provided the Lunar seismic data and all the ancillary information that made possible the presented analyses.

A14.2 THE CONTEXT

Lunar exploration is relevant not only for scientific reasons (AA.VV., 1975) but for possible future mining activities as well (Chamberlain et al., 1993; Sherwood and Woodcock, 1993).

Active and passive seismic data collected during various Apollo missions (*Active Seismic Experiment*, ASE and *Lunar Seismic Passive Experiment*, LSPE) were analyzed by several authors for the reconstruction of both the near-surface and deep Lunar structure (AA.VV., 1972; Kovach and Watkins, 1973; Watkins and Kovach, 1973; Mark and Sutton, 1975;

Surface Wave Analysis for Near Surface Applications
ISBN 978-0-12-800770-9, http://dx.doi.org/10.1016/B978-0-12-800770-9.15014-9

Chenet et al., 2006) while several laboratory measurements were performed with the aim of obtaining an appropriate framework within which to interpret Lunar data (Tittmann, 1975).

Refraction studies allowed to determine the P-wave velocity profiles of the shallowest materials (Watkins and Kovach, 1973) and the obtained velocities (remarkably low—Figure A14.1) were generally explained in terms of formation mechanism of the surface soft layers and the anhydrate conditions of the Lunar materials.

The regolith is represented by a layer of unconsolidated debris constituted by fine soil (average density around 1.5–1.8 g/cm^3) occasionally including breccia, rocks, and boulders from the bedrock.

In general terms, it is also notable the ringy nature of the Lunar seismic signals determined by very high seismic quality factors Q (low attenuation) related to the absence of liquids and gases in a highly fractured material and, possibly, to vacuum conditions and high-temperature effects (Tittmann, 1975; Tittmann et al., 1979).

The current case study briefly reports the reprocessing of some data acquired in the framework of both the active and passive experiments of the Apollo 16 mission (April 16–27, 1972) (see Dal Moro, 2013 and, for further details, a near-future wider paper). In particular, a joint analysis of Rayleigh-wave dispersion (ASE data) and HVSR (LSPE data) is performed also considering the effects of attenuation and with the aim of retrieving the local shear-wave velocity profile.

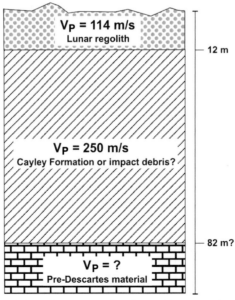

Figure A14.1 V$_P$ model for the Apollo 16 site. *After the Apollo 16—Preliminary Science Report, AA.VV. (1972).*

A14.3 APPROACHING THE ANALYSIS

In order to determine a robust V_S profile, a joint analysis procedure based on Rayleigh-wave dispersion (from active acquisitions) and HVSR (from the passive experiment) was implemented.

Rayleigh-wave dispersion was depicted by means of the *Multiple Filter Analysis* technique (Dziewonsky et al., 1969; Herrmann, 2003) for three reasons:

1. It can be used to extract the dispersive properties of the materials (defined in terms of group velocities) while considering a single seismic trace.
2. If compared with methodologies that rely on the phase velocities, to some degree group velocities appear more sensitive to depict subsurface V_S variations (Luo et al., 2011).
3. Due to the highly ringy and scattered nature of Lunar seismic data, probably related to a highly heterogeneous regolith consisting of powderlike materials with massive boulders and overall low attenuation, *multichannel analysis of surface waves* (a technique based on trace correlation) cannot be easily adopted; Performed tests—not shown here for the sake of brevity—seem to confirm that.

About the first point, it must be also considered that seismic traces of the Lunar missions are often clipped due to the limited dynamic range (5 bit A/D converter) of the pioneering equipment used during the Apollo missions.

Since mode identification can be particularly tricky when analyzing group-velocity spectra (see Chapter 3), the Rayleigh-wave dispersion was analyzed while considering the *Full Velocity Spectrum* (see Chapter 6).

Since the ASE datasets necessarily consider small offsets (see AA.VV., 1972) it is clear that group-velocity spectra of the Lunar ASE data have a limited investigation depth. For this reason (and to further constrain the inversion process), HVSR was also considered (Mark and Sutton, 1975; Arai and Tokimatsu, 2005).

According to the principia illustrated in Chapter 5, the joint inversion of Rayleigh-wave dispersion and HVSR was considered in the framework of the Pareto dominance criterion. In this bi-objective approach, the two cost functions (the group-velocity spectrum and the HVSR misfits) are kept separated and the model distribution in the objective space is used to validate the overall inversion procedure and results (e.g., Dal Moro and Ferigo, 2011).

A14.4 THE APOLLO 16 DATASET

The Apollo 16 landing site was selected in order to investigate the *Descartes* and the *Cayley* Formations present on the *Descartes* highlands.

Figure A14.2 reports a seismic trace (from the Apollo 16 ASE) and its spectrogram, putting in evidence the Rayleigh-wave arrivals (and the highly scattering nature of the data/site). Thanks to the ASE data, it was possible to determine the

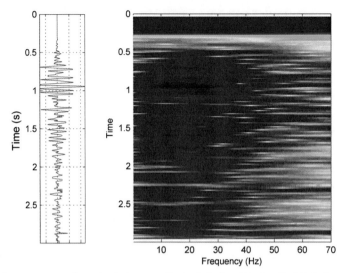

Figure A14.2 Spectrogram of a seismic trace collected during the Apollo 16 Active Seismic Experiment: arrival times of surface waves range mainly in the 0.5 and 1.1 s time window. Considering the offset of the considered trace (32 m), these correspond to group velocities of about 30—60 m/s (see group-velocity spectrum in Figure A14.3).

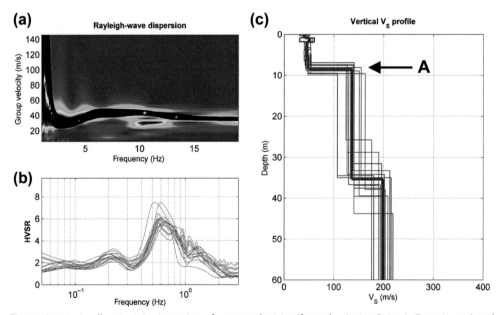

Figure A14.3 Apollo 16—joint inversion of group velocities (from the Active Seismic Experiment) and horizontal-to-vertical spectral ratio (HVSR) (from the Lunar Seismic Passive Experiment) via *multiobjective evolutionary algorithm* (Pareto-front analysis): (a) observed and synthetic group-velocity spectra (background colors represent the observed data while the overlying contour lines correspond to the "best" identified model); (b) observed (green line) and synthetic HVSR curves (shown the curves of all the Pareto-front models); (c) V_S profiles of the Pareto-front models (the "A" horizon indicates the depth of the Lunar regolith).

group-velocity spectrum reported in Figure A14.3(a) (the background color velocity spectrum). About the HVSR, it must be considered that natural seismic events on the Moon belong fundamentally to three main categories: meteoroid impacts, shallow moonquakes, and deep moonquakes. The first two kinds have somehow similar features (AA.VV., 1972; Dal Moro, 2014a) but are not as frequent as the deep events which, because of their low amplitude (the reader must keep in mind that the equipment dates back to 1972), cannot be easily considered for HVSR computation. For the present study, we considered the HVSR associated to a meteoroid impact occurred on May 13, 1972 (Dal Moro, 2013).

Results of the joint inversion procedure (see Chapters 5 and 6) are reported in Figure A14.3 (the contact between the superficial soft Lunar regolith and a deeper material appears very clear—see Horizon "A").

Some remarks about the role of attenuation while computing HVSR is definitely necessary. Together with the observed HVSR (green line), Figure A14.4 reports two synthetic HVSR curves. The first one (blue line) relates to one of the Pareto-front models reported in Figure A14.3, while the second one (magenta dotted line) refers to the same V_S model after having modified the Q_S values (keeping all the other parameters as in previous model) to make them closer to those we would have for similar terrestrial soils/rocks. In general terms in fact, Q_S (and Q_P) values are related to V_S (and V_P) values. Massive rocks such as granite will have very high V_S (and V_P) values as well as very high Q_S (and Q_P). On the other hand, very soft sediments such as peats or sands will have low V_S and Q_S and, in general terms, it is therefore possible to imagine a kind of $Q_S = V_S/k$ rule of thumb (see also Chapter 3).

The performed analyses show that for the Lunar data such k factor is extremely small when compared with the terrestrial case: roughly speaking, while for this latter

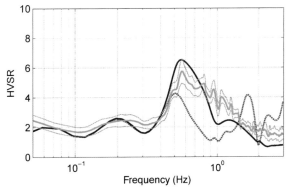

Figure A14.4 Role of attenuation on Lunar horizontal-to-vertical spectral ratio (HVSR). Green line: observed HVSR; blue line: HVSR of one of the Pareto-front models (see Figure A14.3); magenta line: HVSR of the same V_S model but now considering Q_S values typical of analogous (in terms of V_S values) terrestrial materials.

case k factor is in the order of magnitude of 10, for the Lunar data k seems smaller by 2–3 order of magnitudes. In fact, Q_S values for terrestrial soft sediments with V_S values around 100 m/s (or less) are typically in the 5–10 range, while in the Lunar case the Q_S values of the shallowest layers (characterized by V_S values between 40 and 100 m/s) are approximately in the range of 100–200 (further details will be published in a future article currently in progress).

Incidentally, this fact is also confirmed by laboratory measurements and eventually mirrors in the seismological behavior of our satellite (see next paragraph).

A14.5 SOME FINAL REMARKS

Lunar seismic data were analyzed with the main goal of retrieving a robust V_S vertical profile of the shallow layers at the Apollo 16 landing site. As recommended throughout the book, a joint approach (in this case represented by the analysis of Rayleigh-wave group velocities and HVSR) was adopted.

The presented analyses clearly support the idea that a very-well defined horizon is present below the Lunar regolith at a depth of about 10 m (compare the V_P model in Figure A14.1 and the horizon "A" in Figure A14.3(c)). While the regolith can be described by a very-soft powder stratum (average $V_P \approx 110$ m/s, average $V_S = 30{-}60$ m/s, Poisson value $= 0.33{-}0.45$), the underlying material (often indicated as *mega-regolith*— $V_P \approx 250$ m/s, $V_S \approx 120$ m/s, Poisson value $= 0.27$) is typically interpreted as a highly brecciated material whose elastic properties are strongly affected by thermal cracking.

The extremely poor geomechanical properties of the Lunar regolith are easily explained by the peculiar mechanisms involved in its creation. Such processes (often referred to as *impact gardening* or *space weathering*) are fundamentally caused by the absence of an atmosphere with the consequent heavy meteoric rain that constantly hits the Lunar surface and the extreme temperature oscillations that crack the Lunar rocks (temperature approximately ranges from $-150\ ^\circ$C on the dark side to $+120\ ^\circ$C on the sunlit surface).

The two pictures in Figure A14.5 clearly show the effects of even micrometeorites on the Lunar surface which, consequently, crumble. The very poor geomechanical properties of the Lunar surface was in fact a serious concern for the NASA staff in charge of all the operations involved in the Apollo module landing and we can conjecture that everything would have been different if the Lunar gravity would be higher (in that case the carrying capacity of the Lunar soil might be not sufficient to hold on a spacecraft).

Presented analyses seem also to confirm very high Q values also for the shallowest materials (compare laboratory and field data analyses reported in Tittmann (1975) and Tittmann et al. (1972, 1975, 1978, 1979)).

Such high Q values (low attenuation) are also responsible for the abnormal (compared to the terrestrial case) length of even small moonquakes. Figure A14.6 shows the

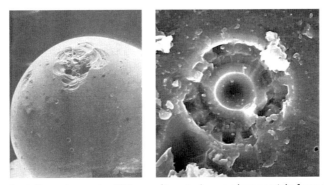

Figure A14.5 On the left, a microcrater (0.3 mm diameter) on a glass particle from the Lunar regolith *(from Guest and Greeley (1977))*; on the right, a 10-μ diameter microcrater. *(from http://stardust.jpl.nasa. gov).*

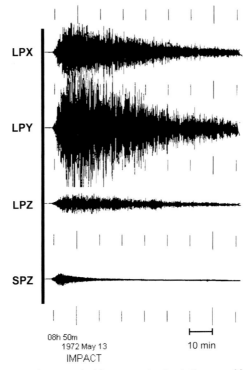

Figure A14.6 Impact moonquake recorded by two seismic stations on May 13, 1972: LPX, LPY, and LPZ relate to a 3-component station, while SPZ to a single (vertical component) one. *After Nakamura et al. (1974).*

recording of an impact moonquake on two stations: the length of the event (more than 1.5 h) is quite remarkable if we consider the limited amount of energy involved.

It can be finally underlined that since the acquisition of geophysical data on extraterrestrial bodies must necessarily rely on light equipment and follow simple procedures, the employment of techniques like the ones presented here can result significantly efficient also for the near-future exploration of Mars.

REFERENCES

AA.VV., SESAME, 2005. Guidelines for the Implementation of the H/V Spectral Ratio Technique on Ambient Vibrations Measurements, Processing and Interpretation open file. http://www.geo.uib.no/seismo/software/jsesame/SESAME-HV-User-Guidelines.pdf, 62 pp. (last access April 2014).

AA.VV., 1972. Apollo 16-Preliminary Science Report (NASA SP 315), National Aeronautics and Space Administration, Washington, D.C., pp. 626.

AA.VV., 1975. The Apollo Program Summary Report, National Aeronautics and Space Administration (NASA) – Houston, Texas, pp. 673.

Albarello, D., Lunedei, E., 2010. Alternative interpretations of horizontal to vertical spectral ratios of ambient vibrations: new insights from theoretical modeling. Bull. Earthquake Eng. 8, 519–534.

Albarello, D., 2006. Possible Effects of Regional Meteoclimatic Conditions on HVSR. NATO SfP 980857-II° Intermediate Meeting, Dubrovnik 25-27 November 2006, open file. nato.gfz.hr/Presentation_Albarello.pdf.

Ali, M.Y., Berteussen, K.A., Small, J., Barkat, B., Pahlevi, O., 2009. Results from a low frequency passive seismic experiment over an oilfield in Abu Dhabi. First Break 27, 91–97.

Arai, H., Tokimatsu, K., 2005. S-Wave velocity profiling by joint inversion of microtremor dispersion curve and horizontal-to-vertical (H/V) spectrum. Bull. Seism. Soc. Am. 95, 1766–1778.

Arai, H., Tokimatsu, K., 2004. S-wave velocity profiling by inversion of microtremor H/V spectrum. Bull. Seism. Soc. Am 94, 53–63.

Asten, M.W., 1978. Geological control on the three-component spectra of Rayleigh-wave microseisms. Bull. Seism. Soc. Am. 68, 1623–1636.

Asten, M.W., Henstridge, J.D., 1984. Array estimators and the use of microseisms for reconnaissance of sedimentary basins. Geophysics 49, 1828–1837.

Baker, G.S., 1999. In: Young, R.A. (Ed.), Processing Near-Surface Seismic-Reflection Data: A Primer. Society of Exploration Geophysicists Publications, p. 78.

Barton, N., 2006. Rock Quality, Seismic Velocity, Attenuation and Anisotropy. Taylor & Francis, 756 pp.

Bonnefoy-Claudet, S., Köhler, A., Cornou, C., Wathelet, M., Bard, P.-Y., 2008. Effects of Love Waves on Microtremor H/V Ratio. Bull. Seism. Soc. Am. 98, 288–300.

Bonnefoy-Claudet, S., Cotton, F., Bard, P.-Y., 2006. The nature of noise wavefield and its applications for site effects studies: A literature review. Earth-Sci. Rev. 79, 205–227.

Carcione, J.M., 1992. Modeling anelastic singular Surface waves in the Earth. Geophysics 57, 781–792.

Cary, P.W., Zhang, C., 2009. Ground roll attenuation with adaptive eigenimage filtering. In: Proceedings SEG, Houston 2009 International Exposition and Annual Meeting, pp. 3302–3306.

Cessaro, R.K., 1994. Sources of Primary and Secondary Microseisms. Bull. Seism. Soc. Am. 84, 142–148.

Cha, Y.H., Kang, J.S., Jo, C.H., 2006. Application of linear-array microtremor surveys for rock mass classification in urban tunnel design. Explor. Geophys. 37, 108–113.

Chamberlain, P.G., Taylor, L.A., Podnieks, E.R., Miller, R.J., 1993. A Review of Possible Mining Applications in Space. In: Lewis, John S., Matthews, Mildred S., Guerrieri, Mary L. (Eds.), Resources of near-earth space. Space Science Series. The University of Arizona Press, 1993, Tucson, London, pp. 51–68.

Chenet, H., Lognonné, P., Wieczorek, M., Mizutani, H., 2006. Lateral variations of lunar crustal thickness from the Apollo seismic data set. Earth Planet. Sci. Lett. 243, 1–14.

Cho, I., Senna, S., Fujiwara, H., 2013. Miniature array analysis of microtremors. Geophysics 78, KS13–KS23.

Coello Coello, C.A., 2003. Guest editorial: special issue on evolutionary multiobjective optimization. IEEE Trans. Evol. Comput. 7, 97–99.

Coello Coello, C.A., 2002. Evolutionary Multiobjective Optimization: Past, Present and Future open file. http://www.cs.cinvestav.mx/~EVOCINV/download/tutorial-moea.pdf (accessed April 2014).

Constable, S.C., Parker, R.L., Constable, C.G., 1987. Occam's inversion: A practical algorithm for generating smooth models from electromagnetic sounding data. Geophysics 52, 289–300.

Dal Moro, G., 2014a. Joint Analysis of Lunar Seismic Data from the Missions Apollo 14 and 16 (in progress).

Dal Moro, G., 2014b. An holistic and Efficient Approach for Surface-Wave Acquisition and Analysis (in progress).

Dal Moro, G., Marques Moura, R.M., 2014. Multi-component acquisition and joint analysis of surface waves, submitted to J. Appl. Geophysics.

Dal Moro, G., Coviello, V., Del Carlo, G., 2014. Shear-wave velocity reconstruction via unconventional joint analysis of seismic data: a case study in the light of some theoretical aspects. IAEG (International Association for Engineering Geology and the Environment) XII CONGRESS – Turin, September 15–19, 2014. In: Lollino, G., et al. (Eds.), Engineering Geology for Society and Territory, 5. Springer International Publishing.

Dal Moro, G., Keller, L., 2013. Unambiguous determination of the V_S profile via joint analysis of multi-component active and passive seismic data. Near Surface 2013. In: Proceedings of the 19th European Meeting of Environmental and Engineering Geophysics, Bochum, Germany, September 9–11, 2013. EAGE.

Dal Moro, G., 2013. Joint analysis of lunar Surface waves: the Apollo 16 dataset. Near Surface 2013. In: Proceedings of the 19th European Meeting of Environmental and Engineering Geophysics, Bochum, Germany, September 9–11, 2013. EAGE.

Dal Moro, G., Ferigo, F., 2011. Joint analysis of Rayleigh and Love wave dispersion for near-surface studies: issues, criteria and improvements. J. Appl. Geophys. 75, 573–589.

Dal Moro, G., 2011. Some aspects about surface wave and HVSR analyses: a short overview and a case study. Boll. Geofis. Teor. Appl. (BGTA) 52, 241–259.

Dal Moro, G., 2010. Insights on surface wave dispersion and HVSR: joint analysis via pareto optimality. J. Appl. Geophys. 72, 29–140.

Dal Moro, G., 2008. V_S and V_P vertical profiling and Poisson ratio estimation via joint inversion of Rayleigh waves and refraction travel times by means of bi-objective evolutionary algorithm. J. Appl. Geophys. 66, 15–24.

Dal Moro, G., Pipan, M., Gabrielli, P., 2007. Rayleigh wave dispersion curve inversion via genetic algorithms and posterior probability density evaluation. J. Appl. Geophys. 61, 39–55.

Dal Moro, G., Pipan, M., 2007. Joint inversion of surface wave dispersion curves and reflection travel times via multi-objective evolutionary algorithms. J. Appl. Geophys. 61, 56–81.

Dal Moro, G., Pipan, M., Forte, E., Finetti, I., 2003. Determination of Rayleigh wave dispersion curves for near surface applications in unconsolidated sediments. Proceedings SEG. In: 73rd Annual Meeting, Dallas, Texas, October 26–31, 2003. Society of Exploration Geophysicists, pp. 1247–1250.

Dunkin, J.W., 1965. Computation of modal solutions in layered, elastic media at high frequencies. Bull. Seism. Soc. Am. 55, 335–358.

Dziewonsky, A., Bloch, S., Landisman, N., 1969. A technique for the analysis of transient seismic signals. Bull. Seism. Soc. Am. 59, 427–444.

Evison, F.F., Orr, R.H., Ingham, C.E., 1959. Thickness of the earth's crust in Antarctica. Nature 183, 306–308.

Fah, D., Kind, F., Giardini, D., 2001. A theoretical investigation of average H/V ratios. Geophys. J. Int. 145, 535–549.

Fonseca, C.M., Fleming, P.J., 1993. Genetic algorithms for multiobjective optimization: formulation, discussion and generalization. In: Proceeding of the fifth International Conference on Genetic Algorithms, San Mateo, CA (USA). Morgan-Kauffman, pp. 416–423.

Forbriger, T., 2003a. Inversion of shallow-seismic wavefields. Part I: wavefield transformation. Geophys. J. Int. 153, 719–734.

Forbriger, T., 2003b. Inversion of shallow-seismic wavefields. Part II: inferring subsurface properties from wavefield transforms. Geophys. J. Int. 153, 735–752.

Foti, S., Lancellotta, R., Sambuelli, L., Socco, L., 2000. Notes on fk analysis of surface waves. Ann. Geofis. 43, 1199–1209.

Gaherty, J.B., 2004. A surface wave analysis of seismic anisotropy beneath eastern north America. Geophys. J. Int. 158, 1053–1066.

Gedge, M., Hill, M., 2012. Acoustofluidics 17: theory and applications of surface acoustic wave devices for particle manipulation. Lab Chip 12, 2998–3007.

Goldberg, D.E., 1989. Genetic Algorithms in Search, Optimization, and Machine Learning. Addison-Wesley Publishing Company Inc.

Groos, L., 2013. 2D Full Waveform Inversion of Shallow Seismic Rayleigh Waves. Dissertation thesis. Karlsruher Instituts für Technologie, 156 pp.

Groos, L., Schäfer, M., Forbriger, T., Bohlen, T., 2013. Comparison of 1D conventional and 2D full waveform inversion of recorded shallow seismic Rayleigh waves. Near Surface 2013. In: Proceedings of the 19th European Meeting of Environmental and Engineering Geophysics, Bochum, Germany, September 9–11, 2013. EAGE.

Guest, J.E., Greeley, R., 1977. Geology on the Moon. Taylor & Francis, 220 pp.

Gutenberg, B., 1958. Microseismic. Adv. Geophys. 5, 53–92.

Gutenberg, B., 1924. Dispersion und Extinktion von seismischen Oberflächenwellen und der Aufbau der obersten Erdschichten. Physikalishce Zeitschrift 25, 377–381.

Harris, P., Du, Z., MacGregor, L., Olsen, W., Shu, R., Cooper, R., 2009. Joint interpretation of seismic and CSEM data using well log constraints: an example from the Luva field. First Break 27, 73–81.

Hayashi, K., Suzuki, H., 2004. CMP cross-correlation analysis of multi-channel surface-wave data. Explor. Geophys. 35, 7–13.

Herak, M., 2008. ModelHVSR – a Matlab tool to model horizontal-to-vertical spectral ratio of ambient noise. Comput. Geosci. 34, 1514–1526.

Herrmann, R.B., 2003. Computer Programs in Seismology open file. http://www.eas.slu.edu/People/RBHerrmann/CPS330.html.

Hashash, Y.M.A., Park, D., 2001. Non-linear one-dimensional seismic ground motion propagation in the Mississippi embayment. Eng. Geol. 62, 185–206.

Holland, J.H., 1975. Adaptation in Natural and Artificial Systems. The University of Michigan Press pp. 206.

Ivanov, J., Miller, R.D., Xia, J., Steeples, D., Park, C.B., 2006. Joint analysis of refractions with Surface waves: an inverse refraction-traveltime solution. Geophysics 71, R131–R138.

Ivanov, J., Miller, R.D., Xia, J., Steeples, D.W., Park, C.B., 2005a. The inverse problem of refraction traveltimes, part I: types of geophysical nonuniqueness through minimization. Pure Appl. Geophys. 162 (3), 447–459.

Ivanov, J., Miller, R.D., Xia, J., Steeples, D., 2005b. The inverse problem of refraction traveltimes, part II: quantifying refraction nonuniqueness using a three-layer model. Pure Appl. Geophys. 162, 461–477.

Jørgensen, C.S., Kundu, T., 2002. Measurement of material elastic constants of trabecular bone: a micromechanical analytic study using a 1 GHz acoustic microscope. J. Orthop. Res. 20, 151–158.

Klein, G., Bohlen, T., Theilen, F., Kugler, S., Forbriger, T., 2005. Acquisition and inversion of dispersive seismic waves in shallow marine environments. Mar. Geophys. Res. 26, 387–315.

Köhler, A., Ohrnberger, M., Scherbaum, F., Wathelet, M., Cornou, C., 2007. Assessing the reliability of the modified three-component spatial autocorrelation technique. Geophys. J. Int. 168, 779–796.

Koper, K.D., Seats, K., Benz, H., 2010. On the composition of Earth's short-period seismic noise field. Bull. Seism. Soc. Am. 100, 606–617.

Kovach, R.L., Watkins, J.S., 1973. The structure of the Lunar crust at the Apollo 17 site. Proceedings of the Forth Lunar Science Conference, Supplement 4 Geochim. Cosmochim. Acta 3, 2549–2560.

Kovach, R.L., 1978. Seismic surface waves and crustal and upper mantle structure. Rev. Geophys. 16, 1–13.

Lai, C.G., Rix, G.J., 2002. Solution of the Rayleigh Eigenproblem in viscoelastic media. Bull. Seism. Soc. Am. 92, 2297–2309.

Lai, C.G., Rix, G.J., 1998. Simultaneous Inversion of Rayleigh Phase Velocity and Attenuation for Near-Surface Site Characterization. Georgia Institute of Technology, School of Civil and Environmental Engineering. Report No. GIT-CEE/GEO-98-2, July 1998, pp. 258.

Louie, J.N., 2001. Faster, better: Shear-wave velocity to 100 meters depth from Refraction Microtremor arrays. Bull. Seism. Soc. Am. 91, 347–364.

Louis, S.J., Chen, Q., Pullammanappallil, S., 1999. Seismic velocity inversion with genetic algorithms. In: CEC99, 1999 Congress on Evolutionary Computation, Mayflower Hotel, Washington D.C., July 6–9, 1998, pp. 855–861.

Love, A.E.H., 1911. Some Problems of Geodynamics. Cambridge University Press.

Lunedei, E., Albarello, D., 2009. On the seismic noise wavefield in a weakly dissipative layered Earth. Geophys. J. Int. 177, 1001–1014.

Luo, Y., Xia, J., Xu, Y., Zeng, C., 2011. Analysis of group-velocity dispersion of high-frequency Rayleigh waves for near-surface applications. J. Appl. Geophys. 74, 157–165.

Luo, Y., Xia, J., Miller, R.D., Xu, Y., Liu, J., Liu, Q., 2009. Rayleigh-wave mode separation by high-resolution linear Radon transform. Geophys. J. Int. 179, 254–264.

Luo, Y., Xia, J., Liu, J., Liu, Q., Xu, S., 2007. Joint inversion of high-frequency surface waves with fundamental and higher modes. J. Appl. Geophys. 62, 375–384.

Malagnini, L., Herrmann, R.B., Mercuri, A., Opice, S., Biella, G., de Franco, R., 1997. Shear-wave velocity structure of sediments from the inversion of explosion-induced Rayleigh waves: comparison with cross-hole measurements. Bull. Seism. Soc. Am. 87, 1413–1421.

Malagnini, L., Herrmann, R.B., Biella, G., de Franco, R., 1994. Rayleigh waves in quaternary alluvium from explosive sources: determination of Shear-wave velocity and Q structure. Bull. Seism. Soc. Am. 85, 900–922.

Mark, N., Sutton, G.H., 1975. Lunar shear velocity structure at Apollo sites 12, 14, and 15. J. Geophys. Res. 80, 4932–4938.

McMechan, G.A., Yedlin, M.J., 1981. Analysis of Dispersive waves by wave field transformation. Geophysics 46, 869–874.

Miller, G.F., Pursey, H., 1955. On the partition of energy between elastic waves in a semi-infinite solid. Proc. R. Soc. (London) 233, 55–69.

Mucciarelli, M., Gallipoli, M.R., Arcieri, M., 2003. The stability of the horizontal-to-vertical spectral ratio of triggered noise and earthquake recordings. Bull. Seism. Soc. Am. 93, 1407–1412.

Nakamura, Y., 2000. Clear identification of fundamental idea of Nakamura's technique and its applications. In: Proc XII World Conf. Earthquake Engineering, New Zealand. Paper no 2656.

Nakamura, Y., 1996. Realtime Information Systems for Seismic Hazard Mitigation. Quarterly Report of Railway Technical Research Inst. (RTRI), 37, 112–127.

Nakamura, Y., 1989. A Method for Dynamic Characteristics Estimation of Subsurface Using Microtremor on the Ground Surface. Quarterly Report of Railway Technical Research Inst. (RTRI), 30, 25–33.

Nakamura, Y., Dorman, J., Duennebier, F., Ewing, M., Lammelein, D., Latham, G., 1974. High-frequency lunar teleseismic events. In: Proc. Lunar Sci. Conf. 5 Geochim. Cosmochim. Acta 5, 2883–2890.

Natale, M., Nunziata, C., Panza, G.F., August 1–6, 2004. FTAN method for the detailed definition of Vs in urban areas. In: 13th World Conference on Earthquake Engineering, p. 2694. Vancouver, B.C., Canada.

O'Connell, D.R.H., Turner, J.P., 2011. Interferometric Multichannel Analysis of Surface Waves (IMASW). Bull. Seism. Soc. Am. 101, 2122–2141.

Ohori, M., Nobata, A., Wakamatsu, K., 2002. A comparison of ESAC and FK methods of estimating phase velocity using arbitrarily shaped microtremor analysis. Bull. Seism. Soc. Am. 92, 2323–2332.

Okada, H., 2006. Theory of efficient array observations of microtremors with special reference to the SPAC method. Explor. Geophys. 37, 73–85.

Okada, H., 2003. The microseismic survey method. Society of Exploration Geophysicists of Japan. In: Geophysical Monograph, Series No. 12. Society of Exploration Geophysicists, Tulsa.

O'Neill, A., Safani, J., Matsuoka, T., Shiraishi, K., 2006. Rapid shear wave velocity imaging with seismic landstreamers and surface wave inversion. Explor. Geophys. 37, 292–306.

O'Neill, A., Matsuoka, T., 2005. Dominant higher Surface-wave modes and possible inversion pitfalls. J. Environ. Eng. Geophys. 10, 185–201.

O'Neill, A., Matsuoka, T., Tsukada, K., 2004. Some pitfalls associated with dominant highermode surface-wave inversion. In: Near Surface 2004–10th European Meeting of Environmental and Engineering Geophysics, Utrecht, The Netherlands.

O'Neill, A., Dentith, M., List, R., 2003. Full-waveform P-SV reflectivity inversion of surface waves for shallow engineering applications. Explor. Geophys. 34, 158–173.

Panza, G.F., 1989. Attenuation measurements by multimode synthetic seismograms. In: Cassinis, R., Nolet, G., Panza, G.F. (Eds.), Digital Seismology and Fine Modeling of the Lithospehere. Plenum Publishing Corporation, pp. 79–115.

Park, C.B., Xia, J., Miller, R.D., September 13–18, 1998. Imaging dispersion curves of surface waves on multichannel record. In: Proceedings SEG (Society of Exploration Geophysicists) 2003, 68th Annual Meeting, New Orleans, Louisiana, pp. 1377–1380.

Parolai, S., Galiana-Merino, J.J., 2006. Effect of transient seismic noise on estimates of H/V spectral ratios. Bull. Seism. Soc. Am. 96, 228–236.

Parolai, S., Picozzi, M., Strollo, A., Pilz, M., Di Giacomo, D., Liss, B., Bindi, D., 2009. Are transients carrying useful information for estimating H/V spectral ratios? In: Mucciarelli, M., Herak, M., Cassidy, J. (Eds.), Increasing Seismic Safety by Combining Engineering Technologies and Seismological Data. Springer.

Pedersen, H.A., Bruneton, M., Maupin, V., The SVEKALAPKO Seismic Tomography Working group, 2006. Lithospheric and sublithospheric anisotropy beneath the Baltic shield from surface-wave array analysis. Earth Planet. Sci. Lett. 244, 590–605.

Pedersen, H.A., Mars, J.I., Amblard, P.-O., 2003. Improving surface-wave group velocity measurements by energy reassignment. Geophysics 68, 677–684.

Picozzi, M., Albarello, D., 2007. Combining genetic and linearized algorithms for a two-step joint inversion of Rayleigh wave dispersion and H/V spectral ratio curves. Geophys. J. Int. 169, 189–200.

Poggi, V., Fäh, D., 2010. Estimating Rayleigh wave particle motion from three-component array analysis of ambient vibrations. Geophys. J. Int. 180, 251–267.

Prasad, M., 2002. Acoustic measurements in unconsolidated sands at low effective pressure and overpressure detection. Geophysics 67, 405–412.

Prodehl, C., Kennett, B., Artemieva, I.M., Thybo, H., 2013. 100 years of seismic research on the Moho. Tectonophysics 609, 9–44.

Rayleigh, J.W.S., 1885. On waves propagated along the plane surface of an elastic solid. Proc. London Math. Soc. 17, 4–11.

Reeves, C.R., Rowe, J.E., 2003. Genetic Algorithms – Principles and Perspectives. Kluwer Academic Publisher, Norwell.

Ritzwoller, M.H., Lavshin, A.L., 2003. Estimating shallow shear velocities with marine multicomponent seismic data. Geophysics 67, 1991–2004.

Rix, G.J., Lai, C.G., Spang Jr, A.W., 2000. In situ measurement of damping ratio using Surface waves. J. Geothec. Geoenviron. Eng. 126, 472–480.

Rodríguez-Castellanos, A., Sánchez-Sesma, F.J., Luzón, F., Martin, R., 2006. Multiple scattering of elastic waves by subsurface fractures and cavities. Bull. Seism. Soc. Am. 96, 1359–1374.

Roesset, J.M., 1998. Nondestructive dynamic testing of soils and pavements. Tamkang J. Sci. Eng. 1, 61–81.

Sachse, W., Pao, Y.H., 1978. On the determination of phase and group velocities of dispersive waves in solids. J. Appl. Phys. 49, 4320–4327.

Safani, J., O'Neill, A., Matsuoka, T., Sanada, Y., 2005. Applications of Love wave dispersion for improved Shear-wave velocity imaging. J. Environ. Eng. Geophys. 10, 135–150.

Scales, J.A., Smith, M.L., Treitel, S., 2001. Introductory Geophysical Inverse Theory. open file. Samizdat Press, 193 pp. http://samizdat.mines.edu.

Scholte, J.G., 1947. The range and existence of Rayleigh and Stoneley waves. Geophys. J. Int. 5, 120–126.

Sherwood, B., Woodcock, G.R., 1993. Cost and benefits of Lunar oxygen: economics, engineering and operations. In: Lewis, John S., Matthews, Mildred S., Guerrieri, Mary L. (Eds.), Resources of Near-Earth Space. The University of Arizona Press, pp. 199–227. Space Science Series.

Sen, M.K., Stoffa, P.L., 1992. Rapid sampling of model space using genetic algorithms: examples from seismic waveform inversion. Geophys. J. Int. 108, 281–292.

Shtivelman, V., 2002. Surface wave sections as a tool for imaging subsurface inhomogeneities. Eur. J. Environ. Eng. Geophys. 7, 121–138.

Shtivelman, V., 1999. Using surface waves for estimating the Shear-wave velocities in the shallow subsurface onshore and offshore Israel. Eur. J. Environ. Eng. Geophys. 4, 17–36.

Smith, M.L., Scales, J.A., Fischer, T.L., 1992. Global search and genetic algorithms. Geophys. Leading Edge Explor. 11, 22–26.

Snieder, R., 2002. Scattering of surface waves. In: Pike, R., Sabatier, P. (Eds.), Scattering and Inverse Scattering in Pure and Applied Science. Academic Press, San Diego, pp. 562–577.

Stesky, R.M., 1978. Experimental compressional wave velocity measurements in compacting powders under high vacuum – applications to lunar crustal sounding. In: Lunar and Planetary Science Conference, 9th,

Houston, Tex., March 13–17, 1978, Proceedings, vol. 3. Pergamon Press, Inc., New York, pp. 3637–3649 (A79-39253 16-91).

Stevens, J.L., Day, S.M., 1986. Shear velocity logging in slow formations using the Stoneley wave. Geophysics 51, 137–147.

Stoffa, P.L., Sen, M.K., 1991. Nonlinear multiparameter optimisation using genetic algorithms: inversion of plane wave seismograms. Geophysics 56, 1794–1810.

Stokoe, K.H., September 7–9, 2009. Seismic and Laboratory Seismic Measurements in Civil Engineering Applications. Keynote Speech at "Near Surface 2009". EAGE, Dublin, Ireland.

Stokoe, K.H., Santamarina, J.C., 2000. Seismic-wave-based testing in geotechnical engineering. Geotech. Eng. 2000, 1490–1536.

Stokoe II, K.H., Nazarian, S., Rix, G.J., Sanchez-Salinero, I., Sheu, J., Mok, Y., 1988. In situ seismic testing of hard-to-sample soils by Surface wave method. Earthquake Eng. and Soil Dyn. II – Recent adv. in ground-motion eval. ASCE, Park City, 264–277.

Tasic, I., Runovc, R., 2010. How to test the reliability of instruments used in microtremor horizontal-to-vertical spectral ratio measurements. Acta Geotechinca Slovena 2, 17–28.

Tanimoto, T., Ishimaru, S., Alvizuri, C., 2006. Seasonality in particle motion of microseisms. Geophys. J. Int. 166, 253–266.

Tanimoto, T., 1999. Excitation of normal modes by atmospheric turbulence: source of long-period seismic noise. Geophys. J. Int. 136, 395–402.

Telford, W.M., Geldart, L.P., Sheriff, R.E., 1990. Applied Geophysics. Cambridge University Press, 770 pp.

Tittmann, B.R., Nadler, H., Clark, V., Coombe, L., 1979. Seismic Q and velocity at depth. In: Proc. Lunar Sci. Conf. 10th, pp. 2131–2145.

Tittmann, B.R., Nadler, N., Richardson, J.M., Ahlberg, L., 1978. Laboratory measurements of p-wave seismic Q on lunar and analog rocks. In: Proc. Lunar Sci. Conf. 9th, pp. 1209–1234.

Tittmann, B.R., 1975. Lunar rock seismic Q in 3000–5000 range achieved in laboratory. In: Runcorn, K. (Ed.), The Moon – a New Appraisal from Space Missions and Laboratory Analyses. The Royal Society, London, pp. 475–481.

Tittmann, B.R., Curnow, J.M., Housley, R.M., 1975. Internal friction quality factor $Q \geq 3100$ achieved in lunar rock 70215,85. In: Proc. Lunar Sci. Conf. 6th, pp. 3217–3226.

Tittmann, B.R., Adbel-Gawad, M., Housley, R.M., 1972. Elastic velocity and Q-factor measurements on Apollo12, 14 and 15 rocks. In: Proc. Lunar Sci. Conf. 3rd, pp. 2565–2575.

Tokimatsu, K., Tamura, S., Kojima, H., 1992. Effects of multiple modes on Rayleigh wave dispersion characteristics. J. Geotech. Eng. ASCE 118 (10), 1529–1543.

Tonn, R., 1991. The determination of the seismic quality factor Q from VSP data: a comparison of different computational methods. Geophys. Prospect. 39, 1–27.

Tselentis, G.-A., 1998. Intrinsic and Scattering Seismic Attenuation in W. Greece. Pure Appl. Geophys. 153, 703–712.

Van Veldhuizen, D.A., Lamont, G.B., 1998a. Multiobjective Evolutionary Algorithms: a History and Analysis. Air Force Institute of Technology. Technical Report TR-98–03, 88.

Van Veldhuizen, D.A., Lamont, G.B., 1998b. Evolutionary Computation and Convergence to a Pareto Front. In: Koza, John R. (Ed.), Late Breaking Papers at the Genetic Programming 1998 Conference. Stanford University, pp. 221–228.

Van Veldhuizen, D.A., Lamont, G.B., 2000. Multiobjective evolutionary algorithms: analyzing the State-of-the-Art. Evol. Comput. 8, 125–147.

Viktorov, I.A., 1967. Rayleigh and Lamb Waves: Physical Theory and Applications. Plenum Press, New York, 154 pp.

Watkins, J.S., Kovach, R.L., 1973. Seismic investigation of the Lunar regolith. Proceedings of the Forth Lunar Science Conference, Supplement 4 Geochim. Cosmochim. Acta 3, 2561–2574.

Wen, K.-L., Beresnev, I.A., Yeh, Y.T., 1994. Nonlinear soil amplification inferred from downhole strong seismic motion data. Geophys. Res. Lett. 21, 2625–2628.

West, M., Menke, W., 2001. Fluid-induced changes in shear velocity from Surface waves. In: Symposium on the Application of Geophysics to Engineering and Environmental Problems (SAGEEP), pp. 21–28.

White, R.E., 1992. The accuracy of estimating Q from seismic data. Geophysics 57, 1508–1511.

Winsborrow, G., Huwsa, D.G., Muyzertb, E., 2003. Acquisition and inversion of Love wave data to measure the lateral variability of geo-acoustic properties of marine sediments. J. Appl. Geophys. 54, 71–84.

Xia, J., Miller, R.D., Park, C.B., Tian, G., 2002. Determining Q of near-surface materials from Rayleigh waves. J. Appl. Geophys. 51, 121–129.

Xia, J., Miller, R.D., Park, C.B., 1999. Estimation of near-surface Shear-wave velocity by inversion of Rayleigh waves. Geophysics 64, 691–700.

Yamanaka, H., 2005. Comparison of performance of heuristic search methods for phase velocity inversion in shallow Surface wave method. J. Environ. Eng. Geophys. 10, 163–173.

Yapo, P.O., Gupta, H.V., Sorooshian, S., 1998. Multi-objective global optimization for hydrologic models. J. Hydrol. 204, 83–97.

Yilmaz, O., 1987. Seismic Data Processing. Soc. Explor. Geophys., Tulsa.

Zhang, S.X., Chan, L.S., 2003. Possibile effects of misidentified mode number on Rayleigh wave inversion. J. Appl. Geophys. 53, 17–29.

Zitzler, E., Thiele, L., 1999. Multiobjective evolutionary algorithms: a comparative case study and the strength pareto approach. IEEE Trans. Evol. Comput. 3, 257–271.

INDEX

Note: Page numbers with "f" denote figures "t" tables and "b" boxes.

Printed in the United States
By Bookmasters